ALLEN CARR

SMART PHONE DUMB PHONE

ALLEN CARR

with John Dicey

SMART PHONE

DUMB PHONE

Free yourself from digital adiction

SIRIUS

To Jackie, Emily, & Harry

and to

Tim Glynne-Jones & Nigel Matheson for their amazing
contribution in making this book happen

SIRIUS

This edition published in 2019 by Sirius Publishing, a division of
Arcturus Publishing Limited,
26/27 Bickels Yard, 151–153 Bermondsey Street,
London SE1 3HA

ISBN: 978-1-78950-483-5
AD005774US

Printed in the US

ALLEN CARR

Allen Carr was a chain-smoker for over 30 years. In 1983, after countless failed attempts to quit, he went from 60 to 100 cigarettes a day to zero without suffering withdrawal pangs, without using willpower, and without putting on weight. He realized that he had discovered what the world had been waiting for—the easy way to stop smoking—and embarked on a mission to help cure the world's smokers.

As a result of the phenomenal success of his method, he gained an international reputation as the world's leading expert on stopping smoking and his network of centers now spans the globe. His first book, *Allen Carr's Easy Way to Stop Smoking*, has sold over 12 million copies, remains a global bestseller, and has been published in more than 40 different languages. Hundreds of thousands of smokers have successfully quit at Allen Carr's Easyway centers where, with a success rate of over 90 percent, it's guaranteed you'll find it easy to stop or your money back.

Allen Carr's brilliant Easyway method has been successfully applied to weight control, alcohol, debt, refined sugar, and a host of other addictions and issues.

For more information about Allen Carr's Easyway, please
visit **www.allencarr.com**

Allen Carr's Easyway

The key that will set you free

CONTENTS

INTRODUCTION
BY JOHN DICEY, CO-AUTHOR

- Do you find it impossible to sit in a train, in a bar with friends, or even watch TV without repeatedly checking your phone?

- Do you take some form of comfort from unlocking your phone, even if it's for no specific purpose?

- Do you spend hours each day gaming online?

- Do you fear that you'll miss out if you don't check your social media every few minutes and start feeling upset or uncomfortable if you can't "check in?"

These are all indications of digital addiction or tech addiction. It's a condition that is becoming increasingly common, yet society has been slow in recognizing the threat it poses. With our ever-increasing, all-encompassing dependence on digital devices have come unprecedented levels of stress, isolation, procrastination, sleep issues, and inertia.

Smartphones have been deliberately designed to addict us. That's not to say that the brains behind them set out to cause us harm. The elements that make smartphones addictive are exactly the same elements that make them easy and enjoyable to use, intuitive, and extremely useful. Used efficiently, they can help us control and enhance our lives, assisting us in an endless number of ways in such a significant and effective way; it would have been the stuff of science fiction at the start of the millennium.

Yet, when misused, they control us rather than us controlling them. They nag and disturb us, and demand our attention when we are doing something else. As a result, we gradually fall into the trap of giving in to them and end up doing what they want, when they want. It's all about brain chemistry. Phones are deliberately set up so we keep coming back to them. They alert us, nudge us, and interrupt us no matter what else we might be attempting to focus on or who we might be trying to engage with. They constantly demand our attention.

The average adult spends nearly ten hours a day looking at digital screens: phones, tablets, laptops, desktops, TVs. We have come to regard this as normal, but for many people the digital world they are living in has become obsessive and compulsive, taking over real life, and posing serious health risks. Not just in terms of mental health and the destruction of relationships but issues such as obesity, sleep deprivation, and serious eye strain have all been linked to excessive screen use. Of course, you know all this already—that's why you're reading this book—so rest assured we have no intention of boring you with endless details of the harm that digital addiction has caused you and the fate that will befall you if you fail to escape the "tech trap."

The fact is, you know that digital technology is remolding our brains and turning us into addicts, with the same telltale symptoms as more established addictions:

• inability to control use

• interference with other aspects of life (work and relationships)

• loss of interest in other activities (sports, sex, socializing in person)

- secrecy and deceit

- impatience, irritation, and misery when use is interrupted or curtailed

- inability to concentrate

- realization that all is not well but a devastating inability to change behavior

That last point encapsulates the misery of addiction. Even when we know it's ruining our life and would love to be free of it, escape can seem impossible. Why? Because we think there is only one way out: the hard way.

Whether you're simply concerned about your tech use, struggling to limit it, or in deep despair as a result of digital addiction, this book is for you. The common belief is that addictions cannot be conquered without tremendous willpower, suffering, and deprivation. But this book will show you the beautiful truth: There is another way.

MY STORY

I discovered this for myself 20 years ago, when I went to Allen Carr's center in London, just to satisfy my wife's increasingly desperate requests for me to quit smoking. I had no faith in Allen Carr being able to help me. I smoked 80 a day and had given up all hope of ever being able to quit. I wasn't happy that I smoked, but I believed it was my fate and all attempts to convince me otherwise were pointless.

No one was more surprised than I was (or perhaps my wife) that I walked out of that seminar convinced that I would never smoke again. What I experienced was completely different from my previous attempts to quit by using willpower, substitutes like nicotine gum, and pretty much every quit-smoking method or gimmick known to humanity. By the time I finished Allen Carr's program, I knew I no longer had any desire to smoke, and so I didn't need willpower.

I realized that my fears about life without cigarettes were unfounded: I started enjoying social occasions more and handling stress better than I had as a smoker; there was no feeling of deprivation or missing out; on the contrary, I felt hugely relieved and elated that I was finally free. I felt like I'd been cured of the worst disease I could possibly suffer from.

For a third of a century, Allen had also been a chain-smoker, puffing his way through 60 to 100 cigarettes a day. With the exception of acupuncture, he had tried all the conventional and unconventional methods to quit. Eventually, like me, he gave up even trying to quit, believing "once a smoker, always a smoker," and resigned himself to a premature death. Then he made a discovery that inspired him to try again.

As he described it, "I went overnight from 100 cigarettes a day to zero—without any bad temper or sense of loss, void, or depression. On the contrary, I actually enjoyed the process. I knew I was already a nonsmoker even before I had extinguished my final cigarette and I've never had the slightest urge to smoke since."

This is the outstanding feature of Easyway. Unlike the willpower method, it enables you to conquer your addiction:

• EASILY, IMMEDIATELY, AND PAINLESSLY

• WITHOUT USING WILLPOWER, AIDS, SUBSTITUTES, OR GIMMICKS

- WITHOUT SUFFERING DEPRESSION OR WITHDRAWAL SYMPTOMS

- WITHOUT TURNING TO ALTERNATIVE OBSESSIONS LIKE OVEREATING

So successful has Easyway been that there are now Allen Carr's Easyway centers in more than 150 cities in 50 countries worldwide. Bestselling books based on his method are translated into over 40 languages, with more being added each year. The method has now helped tens of millions of people to quit smoking, alcohol, and other drug addictions, as well as sugar addiction, gambling, overeating, overspending, and fear of flying.

I was so inspired by Allen and what I saw as his miraculous method that I hassled and harangued him and Robin Hayley (now chairman of Allen Carr's Easyway) to let me get involved in their quest to cure the world of smoking. To my good fortune, I succeeded in convincing them. Being trained by Allen and Robin was one of the most rewarding experiences of my life. To be able to count Allen as not only my coach and mentor but also my friend was an amazing honor and privilege.

SHARING THE TRUTH

Over the past 20 years, I have gone on to treat more than 30,000 smokers myself at Allen's original London center and lead the team that has taken his method from Berlin to Bogota, New Zealand to New York, Sydney to Santiago. Tasked by Allen with insuring that his legacy achieves its full potential, we've taken Allen Carr's Easyway from videos to DVD, from seminars to apps, from computer games to audio books, to Online Video Programs and beyond.

Behind this phenomenal success lies one simple truth—a truth that Allen discovered by chance and passed on to tens of millions of people like me. What connects us all is that none of us expected to be changed in the way we were. We were all skeptical, all laboring under the same illusions.

The truth about addictions and compulsive behaviors like smoking and digital addiction is kept hidden from most of us by a carefully orchestrated campaign of cover-ups and falsehoods. The fact is we are all at the mercy of organizations with a vested interest in keeping us hooked, whether it's to a drug, a device, a game, or an app. They have studied the science of addiction and they use it callously to keep us hooked.

Driven by fear of regulation, the tech, gaming, and social media giants have recently engaged in efforts that appear to assist users to be more aware of, and limit, screen time, yet this is akin to a mismatched boxing bout—with a hundred-pound weakling thrown into the ring against a heavyweight champion.

When our use of technology—be it in the form of smartphones, tablets, social media, or gaming—becomes a problem, we attempt to cut down, so we draw on our willpower to hold ourselves back and fight the urge.

Even if our willpower holds out, we still go on feeling a sense of loss, of missing out, and actually crave something we're depriving ourselves of. We never shake off the belief that we are making a sacrifice, "giving up" something that provides us with pleasure or a comfort, or keeps us in the social loop.

Understanding the simple truth and recognizing how it applies to you is the key to escaping the trap of digital addiction and staying permanently free from it.

With straightforward drug addiction, such as nicotine, cocaine, or heroin, Easyway's objective is not only to create a situation whereby complete abstinence is achieved but also to insure that the former addict enjoys a sense of freedom, release, and joy rather than any sense of loss or deprivation. It is the method's effectiveness in achieving that state of mind that has led to it becoming a global phenomenon.

With digital addiction, unless you're planning to completely reject technology for the rest of your life (which even if you planned to go entirely off-grid would present almost insurmountable challenges), our objective is to enable you to eliminate unnecessary, negative, dysfunctional and inappropriate use of technology and replace that with useful, positive, functional and appropriate levels of use.

In other words, we want you no longer to be enslaved, controlled, and used by technology but instead to simply USE IT!

DIGITAL ADDICTION = INAPPROPRIATE OR DYSFUNCTIONAL USE OF TECHNOLOGY

You're quitting inappropriate, dysfunctional use—you're not quitting technology per se (unless you really want to, of course).

This book, like all Easyway books, will help you to see the simple truth. It doesn't rely on guilt, bullying, or scare tactics—as you will learn, all those techniques actually make it harder to quit. Instead, it gives you a structured, easy-to-follow method for overcoming your digital addiction quickly, painlessly, and permanently.

ALLEN'S VOICE

The responsibility for insuring our books are faithful to Allen Carr's original method is mine. It has been suggested to me that I describe myself as the author of the books we've published since Allen passed away. In my view that would be quite wrong.

That's because every new book is written strictly in accordance with Allen Carr's brilliant Easyway method. In our new books, we have updated the method to insure it remains relevant and effective as addiction mechanisms develop and addiction evolves. A good example of this is the necessity for us to include advice and guidance about e-cigarettes and vaping in our stop-smoking books. Of course, we've also developed the method to allow it to be applied to a whole host of other addictions and issues such as alcohol, cocaine, cannabis, debt, sugar addiction, weight issues, fear of flying, and new emerging addictions such as digital addiction. I'm eternally grateful for the huge support provided by our publishers, Arcturus Publishing, in particular by Tim Glynne-Jones and Nigel Matheson.

There is not a word in our books that Allen didn't write, or wouldn't have written, if he was still with us and, for that reason, the updates, anecdotes, and analogies that are not his own work or his own experiences—that were contemporized or added by me—are written clearly in Allen's voice to seamlessly complement the original text and method.

I consider myself privileged to have worked closely with Allen on so many Easyway books while he was alive, gaining insight into how the method could be applied and exploring, and mapping out its future evolution and applicability to other issues and drugs.

I was more than happy to have the responsibility for continuing this vital mission placed on my shoulders by Allen himself. It's a responsibility I accepted with humility and one I take extremely seriously.

The method is as pure, as bright, as adaptable, and as effective as it's ever been, allowing us to apply it to a whole host of addictions and guide those who need help in a simple, relatable, plain-speaking way. I know from happy experience that the benefits of following this method can be life-changing. And now let me pass you into the safest of hands, Allen Carr and his amazing Easyway method.

John Dicey
Global CEO & Senior Therapist, Allen Carr's Easyway

Chapter 1

THE KEY

The rapid development of digital technology and the pressure on us all to keep up has produced a new form of suffering: an unhealthy attachment to digital devices and powerlessness to detach. Behind this lies a very familiar scenario—a carefully planned campaign of brainwashing, designed to trap its victims and control them. We've seen it all before and in Easyway we have the key. Prepare to free yourself from the tyranny of digital addiction.

You're sitting at a table in a favorite restaurant. Around you are friends, who you've been looking forward to catching up with all day. As you all settle in, the phones come out—laid on the table like guns at a Wild West poker game. There's a buzz. You reach for your phone. It's not yours buzzing, but there's a message on the screen that catches your eye.

"Let me just…" you say, and start tapping.

The friends you've looked forward to catching up with fade into the background. You thought it would just take a second to check your messages, but three minutes later you're still scrolling. Your life is inside the device. The real world around you can wait. Checking your messages has become more important. And then you look up from the screen. Your friends are looking at you and shaking their heads.

WHAT ARE YOU DOING?

Embarrassed, you close the phone sheepishly and slip it into your bag. But for the rest of the meal, a little voice is calling you from your bag, dragging your attention away from your friends, nagging at you like a spoiled child, stopping you from relaxing and enjoying their company. And you know that you'll be slipping off to the restroom at the first opportunity for a secret liaison with your phone.

Everybody knows that real-life relationships and face-to-face conversations are more healthy and rewarding than the things we do on our phones, tablets, or whatever other digital devices we use. We know it—but we don't acknowledge it. There is a power in these devices that makes them compulsive. You don't go to a restaurant with the thought of spending the evening on your phone, but when you get there it just takes you over.

Anyone who uses a smartphone, tablet, laptop, fitness tracker, or engages with online gaming will know how these devices and platforms can be both compelling and infuriating at once. This is no accident. The clever people who create the devices, who program them, and those who create the apps that are used on them, want you to keep coming back for more and more and have worked out some fiendish tactics for that very purpose. Frustration is all part of the plan.

They hook you in so that what begins as "Let me just..." always takes longer than you thought. If you check the amount of time you spend on your phone each day, you'll probably find it hard to believe.

Everybody massively underestimates the time these devices take out of their lives.

The fact is that the scene I described at the dinner table probably understates the true situation. More often than not, when you look up from your phone, all you see are the top of people's heads as all those around you are much more engrossed in their phones than in each other.

But digital devices and apps don't just devour your time; they pull the wool over your eyes too. The baffling thing is that this is not news. We joke about how our devices are controlling us. But for a growing number of digital users, it's no joke. The feeling of being controlled goes beyond mere frustration into downright misery, stress, utter confusion, and an inability to focus on the task in hand.

It's a struggle that takes a heavy toll on mental and physical health. Nobody likes to feel controlled. The stress is exhausting, the helplessness damaging to self-esteem, and the compulsion to respond to the little voice, which is often prompted by a "ting," a "buzz," or a telltale vibration leads to secretive behavior that adds to your feelings of alienation and lack of self-respect, threatens your relationships, and leaves you feeling isolated and dissatisfied.

NOW REREAD THE LAST TWO PARAGRAPHS AND REPLACE DIGITAL DEVICES AND APPS WITH NICOTINE OR COCAINE OR HEROIN. THE SYMPTOMS ARE THE SAME

All addictions work in the same way. The more you strive to wrestle back control, the more control you crave, so true satisfaction becomes a more and more distant goal. It's a vicious circle that's as old as the earliest drug.

The good news is that it is not a situation you have to live with for the rest of your life. Next time you find yourself saying, "Let me just…," remember this book rather than your phone and you'll be on your way to freeing yourself from the tyranny of digital addiction.

THE "A"-WORD

Once upon a time, addiction was something that happened to other people—people on the fringes of society: drug addicts, alcoholics, smokers. It was easy to divorce yourself from the term, say "It'll never happen to me," and carry on with your life.

Today, the likelihood is that it *will* happen to you. And the fact that you're reading this book suggests it already has. Some experts reckon that half the world's population are addicted to something. And many of those addictions are behavioral.

In recent years, we've learned that addiction doesn't have to involve a substance, like heroin, alcohol, nicotine, or other drugs; it can be based in a behavior.

Gambling is the obvious example. Problem gamblers display many of the same characteristics as drug addicts: restlessness, obsessiveness, lack of self-control, denial, evasiveness, deceit, irritability, loss of perspective, lack of self-esteem. Advances in scientific research have revealed that drug addictions and behavioral addictions affect the brain in very similar ways, hijacking the natural mental processes to create a perceived dependence on the drug or behavior.

Digital addiction is the "new behavioral addiction on the block." It is a disorder that involves excessive compulsive use of digital devices, causing both physical and mental suffering, not just for the addict, but for their loved ones too.

Later, you will learn how this addiction works and quickly turns a perceived pleasure into absolute slavery. It's actually an ingenious process that twists reality and traps you in a tangle of illusions. But when you see through the illusions—as you will—it all becomes incredibly clear.

Addicts are always craving bigger and bigger doses as the perceived pleasure diminishes. Like all addicts, chronic digital addicts eventually complain of deriving no pleasure at all from their digital use, yet feel powerless to stop it. This is a classic symptom of addiction:

A CHOICE INTENDED TO RELIEVE EMOTIONAL SUFFERING ENDS UP CAUSING IT

It begins small, seemingly under control, but soon grows to the point where it takes you over and starts to ruin your life. You find yourself spending more and more time on devices, utilizing apps, first finding them pretty useful, then finding them essential, before becoming entirely dependent on them. The age-old morning checklist before you leave home has changed from "wallet/purse, keys, coat/umbrella/raincoat (depending on what the weather looks like outside)" to "check emails and my messages from Twitter, check Viber messages, WhatsApp messages and groups, Facebook Messenger, Snapchat, text messages, check Facebook activity, Insta, Twitter, and other social media feeds, check traffic alerts, train/bus alerts, the weather app… wallet/purse, keys, coat/umbrella." No wonder we don't have half a second to kiss or hug our partner or kids before tearing out of the house.

Now quite a few of the above are extremely useful to engage with. Why hurry for a train that, having checked the app, you know to be running a couple of minutes late? Why go without an umbrella when you have a very good idea, from the weather app, that it is likely to rain later? Tackling digital addiction starts with identifying useful tech and ensuring that we use it appropriately.

FEELING UNABLE TO STOP DESPITE THE HARM YOU KNOW IT'S CAUSING IS ANOTHER SURE SIGN OF ADDICTION

As with all addictions, it's the illusion that the behavior provides a genuine pleasure or comfort that keeps you trapped. Smokers suffer from the illusion that cigarettes help them relax. In fact, they do the complete opposite. The same misconceptions are at work when it comes to digital addiction. As you get sucked into engaging with all those different apps and engagements, you're lured into believing that all of these activities are essential for staying in control of your life and being happy—yet the more you do them, the more you feel out of your depth.

Nevertheless, the illusion of pleasure remains.

It's a vicious circle and the longer you go on laboring under the illusion that unlimited use of devices and apps is the answer to achieving happiness and ending anxieties, the more miserable you become. This, in short, is how addiction works. We will explore it in greater detail later on, but at this point, while you're still wondering whether this book really can provide the solution to your problem, it's important to start by acknowledging that your problem is addiction, and there is no shame in that.

Half of all the people in the world are addicted to something.

THE SIMPLE TRUTH

Addiction is inextricably tied up with negative emotions. Stress, depression, fear, heartbreak, worthlessness, loneliness, boredom...

any of these feelings will drive you to seek comfort wherever you think you can find it. If you think you can find it from digital devices, normally via social media and gaming, that's what you'll turn to whenever you feel negative emotions. To be truthful, not many people knowingly do this. It's all part of the trap. We gradually drift into it, normally caught up in the thrill of the first friend request or follow, or a reply to a comment from the original poster on Twitter, or simply someone agreeing wholeheartedly with something you've posted or shared.

That little lift, that small sense of gratification, is all it takes for the chase, the addiction, to begin. All of a sudden the opinion of someone whose real name you don't know, who lives "who knows where," and does "who knows what" for a living, begins to define how you feel about yourself. Eventually your entire mood and demeanor is dictated by the reaction, or nonreaction, of strangers.

That's not to say that real comfort and help and fun cannot be found in carefully selected social media groups. But the less discerning we are about the company we keep online, the more problematic and prone to suffering we become. No doubt you know this already—it's partly why you're reading this book—yet for some reason you can't seem to unhook. The temptation to look at your phone regardless of whether you need to do so for genuine practical reasons, or because you feel an emotional need is too great. It's as if you're being manipulated by a powerful monster.

The purpose of this book is to help you kill that monster and make the escape you know you need to make from the miserable, repetitive cycle of digital addiction, and inappropriate use of technology. Without the tyranny of that monster, you can then deal with your genuine needs with a clear mind.

The solution is simple, yet millions of people like you find it hard to see it for themselves. That monster is quite ingenious. It wraps you up in confusion and blinds you to the simple truth:

THE ONLY WAY TO QUIT AN ADDICTION IS TO STOP FEEDING THE MONSTER

Too simple? Of course, there is more to Easyway than knowing this simple truth, but Easyway works *because* it's much more straightforward than other methods. Addiction *wants* you to think escape is complicated and hard; it *wants* you to lose hope; it *wants* you to lose sight of the obvious: "The only way to quit an addiction is to stop doing it." It's how you go about stopping that matters.

But what exactly does "stopping" mean? Are you supposed to spend the rest of your life without using another digital device?

Rest assured, that is not the case. Taking control of your digital life does not mean avoiding technology completely; on the contrary, it means using it whenever, and wherever, you genuinely need to or want to, and putting down your phone, tablet, or gaming console very happily and without stress when you're done.

The key lies in distinguishing your genuine needs from false needs.

It's safe to assume from the fact that you're reading this book that you do not feel in control of your use of digital devices. Perhaps you've tried to tackle the problem before and found you couldn't. No matter how hard you tried, you seemed to lack the willpower. Now you find yourself looking at devices even when you don't want to and you're powerless to stop. It's become like a constant tic—every few minutes—you've become the digital equivalent of a chain-smoker.

This book will show you two very important truths:

YOU ARE NOT POWERLESS AND YOU DO NOT LACK WILLPOWER

The reason you have failed to overcome your digital addiction up until now is simply because you were following the wrong method.

> **IN CONFIDENCE**
>
> Nobody likes to admit they've lost control. We regard self-control as a cornerstone of civilized behavior. It goes hand in hand with morality, dignity, and courtesy. Loss of control causes feelings of shame and so problems like addiction are swept under the carpet. The person with the problem denies it; the rest try to avoid it.

KEEPING IT ALL TO YOURSELF CREATES A BURDEN THAT ONLY MAKES IT HARDER TO ESCAPE FROM THE TRAP

This is no way to tackle a problem. It makes you feel very alone and compels you to try to conceal it—from others and from yourself. Secretive behavior creates feelings of guilt and shame that deepen your misery.

The first necessity when tackling a problem like digital addiction is to acknowledge that you have it—or that it has you. The fact that you have picked up this book is a good sign that you have made that vital acknowledgement. Next, is to do something about it. The solution is in your hands. More accurately, it's in your mind.

Curing yourself of digital addiction requires you to believe that you *can* cure yourself. Without that belief you are at the mercy of your addiction and your only hope is that you might wake up one morning and the problem will miraculously be gone.

You need to break the cycle of misery. As you read through this book, be honest about your problem and be prepared to open your mind to some truths that may seem hard to accept.

No one is judging you. Nor are you alone. Far from it—digital addiction is fast becoming a global problem, threatening the health and happiness of millions if not billions of people.

As you open your mind to the truths contained in this book, it will become obvious to you.

It will also become apparent that your problem is not down to some failing in your personality. The more you allow yourself to open up and unravel these myths, the more you will understand that you can conquer your digital addiction. And that you can do so without willpower.

A METHOD THAT WORKS

Easyway is a tried and tested way of escaping from addictive traps like digital addiction. It is built on the realization that addiction hijacks your instincts, so that the "fix" you turn to for relief is actually the "poison" that caused your problem in the first place.

That was the revelation that triggered this method. I was a confirmed nicotine addict, choking my way through 60 to 100 cigarettes a day and resigned to a premature death.

I was under the misapprehension that smoking was a habit I had acquired and lacked the willpower to kick. The moment of revelation came when I realized that smoking wasn't just a habit, it was an addiction. That might seem quite obvious in these enlightened times, but back then, more than 35 years ago, it was quite a revelation.

In that moment I saw with extraordinary clarity that my inability to quit smoking was neither a weakness in my character, nor some magical quality in the cigarette. It was the mechanism of addiction that had fooled me into believing that the very thing causing the discomfort and misery was actually curing it. The first cigarette created an empty, insecure feeling as nicotine withdrew from my body; when I smoked my second that slightly uncomfortable feeling seemed to be relieved as nicotine reentered my body. A lifetime chain had started and every cigarette gave the illusion of relief and pleasure. It really was like wearing tight shoes just for the relief of taking them off.

If you've never been addicted to nicotine or been knowingly addicted to a drug, then you probably find that revelation to be a little underwhelming, yet it is the mistaken belief that the "next shot of the drug" relieves the addict's discomfort rather than it being "the previous shot of the drug" that caused it that is the cornerstone of every addictive drug known to humankind.

Every feeling of pleasure, relief, and comfort derived from an addictive drug is founded on that great illusion—of course the addict "feels better" when they take the drug—they're an addict! It's their belief that there is somehow some pleasure or benefit in the drug that makes it so hard for them to get free. A little more about that later.

I don't intend to bore you with too much detail about other addictions, but hopefully you'll understand that sometimes comparing one addiction to another can be useful in illustrating certain points and principles that are essential for you to achieve what you are seeking: FREEDOM.

All of the above led naturally to two indisputable conclusions:

• Drug addiction provides no genuine pleasure or comfort

• Therefore, stopping involves no sacrifice or deprivation

The Easyway method is still hugely successful in helping smokers, and now vapers, all around the world to quit. Once they remove the illusion that they are making a sacrifice by stopping, they find it easy to quit because they don't feel deprived and they are happy to be free.

We knew this method would work for all addictions and we went on to apply it successfully to alcohol, other drugs such as heroin, cocaine, and cannabis, sugar addiction, weight issues, and behavioral addictions like gambling and overspending. The key was that all addictions are mainly a condition of the mind. Easyway is about unraveling the misconceptions that drive you to do something that does you harm in the belief that it will give you pleasure or some kind of benefit. The principle applies to digital addiction just as it does to nicotine addiction, or any other addiction.

If you've ever lost your phone, you will know the panic that grips you when your devices are taken away. But with this method you won't be required to overcome any such panic or pangs. Unlike other methods that rely on willpower to overcome a sense of sacrifice, you will find Easyway is painless. It is also permanent. Having conquered your addiction, you won't be susceptible to falling back into the trap.

Conventional wisdom tells us that behavioral addictions are complex and require immense willpower to overcome. This book will show you that conventional wisdom is misguided. Not only that, it is actually counterproductive when it comes to curing addictions. The truth is that you have the power to overcome digital addiction without any pain or sacrifice, regardless of who you are or what your personal circumstances may be. All you need is an open mind.

If you're skeptical about that, ask yourself one question: Has conventional wisdom regarding digital addiction worked for you until now? If it had, you wouldn't be reading this book.

WHAT THIS BOOK WILL DO FOR YOU

The aims of this book are to:

- Change the way you think about your devices, apps and games

- Enable you to use them mindfully and avoid unhealthy and inappropriate use

- Put a stop to digital addiction—immediately, easily, and painlessly

- Allow you to enjoy the benefits of technology

- Empower you to take control of your life again

It is not a campaign against digital technology. The Internet and all its many offshoots have brought incredible benefits to us and there are a lot more to come. It is pointless and self-destructive to even consider standing in the way of such progress. But if you're already feeling swamped by the digital tidal wave, the promise of connected cars, smart cities, and kitchen appliances that talk to each other and order your food might just drag you under. The aim of this book is to keep you afloat in the sea of digital technology, to enable you to make good use of all the devices and apps at your disposal, and, when you feel they're taking up too much time, or imposing on your life in an unhelpful way, to happily put them down and walk away.

It will tackle the mental process that leads to digital addiction. You will learn to understand what's going on in your mind and why, and then learn to change your mindset so that you no longer feel enslaved by the compulsion to use digital devices in an inappropriate way.

It comes with several assurances:

- You won't be talked down to

- You won't be subjected to scare tactics or gimmicks

- You won't feel deprived

- And by the end, you won't crave your devices, apps or games

By explaining how the addiction trap works and setting out simple, step-by-step instructions to help you get free, this book will show you how to approach digital use for a healthier, happier life.

It will address the feeling of panic that can set in and cloud your judgment when trying to overcome addiction. It will help you to replace deceit, guilt, and embarrassment with openness, honesty, and confidence; it will help you take control where you feel helpless; and it will replace misery with happiness.

There is no need for you to be miserable. You are not "giving up" anything. Nothing bad is going to happen. You will not miss or yearn for anything. You will not feel there is a gap in your life. On the contrary, your life will feel more complete, more balanced, and more relaxed.

THE INSTRUCTIONS

As you read through the book, you will come across a series of instructions. If you miss one of these instructions or fail to follow any of them, the method will not work. If you try to skip ahead and read the book in a different order from which it was written, the method will not work.

Easyway is the key to freeing yourself from the trap you're in and it works like the combination to unlock a safe: If you don't apply all the numbers in the correct order, the lock will not open.

As well as these instructions, I ask you not to make any changes to your digital use at this stage. By the end of the book you will have a completely different attitude, but right now, any attempt to cut down or alter the way you use your digital devices will interfere with this process. That said, it's important that you're able to focus on this book without distractions, so while you are reading it please turn your devices off. If you are reading this book on a device, or listening to it on a device, take care to disable notifications so that you're not disturbed. While you're reading this book, you really don't need to know that Domino's Pizza have a special offer today, or that someone you've never met called Jessie liked your post on Facebook, or that you have a new follower on Twitter or Insta. At all other times please continue to use your devices as normal.

FIRST INSTRUCTION:
FOLLOW ALL THE INSTRUCTIONS

SECOND INSTRUCTION:
DO NOT ALLOW YOUR DEVICES TO INTERRUPT
YOU WHILE YOU ARE READING THIS BOOK

Chapter 2

THE BENEFITS OF DIGITAL TECHNOLOGY

Smartphones, laptops, tablets, desktops, fitness trackers, online gaming... Why do we buy into this stuff? The fact is, digital technology offers us incredible benefits. Our task is to enable you to access those benefits without risking the pitfalls.

With an addiction like smoking, the argument for quitting is easy to win. Smoking is a proven cause of cancer, heart disease, and other life-threatening diseases. Plus, once you understand the addiction you realize that it gives you absolutely no benefit whatsoever. Even hardened smokers accept this argument. The only reason they continue to smoke is because they're hooked on the addictive drug nicotine and they don't know how to get free.

With digital addiction, the argument is not so straightforward because there are numerous undeniable benefits to digital technology. That doesn't mean the method for curing digital addiction has to be different from the method for quitting smoking; in fact, it's exactly the same. But it is important to acknowledge that digital technology is not, in itself, evil, in the way that nicotine or heroin are. Having acknowledged this fact, you will be able to see more clearly the difference between the good, useful apps, social media, and gaming that can benefit you and the bad, junk apps, and inappropriate use of social media and gaming that you need to escape from. So let's take a moment to applaud the incredible benefits of digital technology.

THE DEFINITION OF PROGRESS

The Digital Revolution started somewhere back in the 1950s, when digital circuits started to replace other technologies in functions like counting and record keeping. As with the Agricultural Revolution and Industrial Revolution before it, the Digital Revolution was driven by the natural human instinct for progress.

Progress can be defined in different ways, but as far as these revolutions were concerned, it came down to three factors: the facility to do things

• FASTER

• BIGGER

• CHEAPER

We have a natural desire for speed. We don't like waiting for things. We are not comfortable killing time. And the more things speed up, the higher our expectations of speed become. Just think back to the speed of your home Internet service ten years ago. Imagine how frustrated you would get if it worked at that speed today.

Digital technology hasn't just enabled us to do things faster; it has enabled us to save time spent traveling to places too. Shopping, seeing friends, watching sport—we can do all that and lots more via a screen from the comfort of our own homes or even on the bus on the way home.

The scale at which we do things is also important to us. The Agricultural Revolution was about producing more food from less land using fewer workers; the Industrial Revolution was about mass-producing goods efficiently for sale around the world; and the Digital Revolution is about reaching a wider audience, both socially and commercially, and maximizing our personal and commercial productivity.

Before the Digital Revolution, if you wanted to put on an event in your local town, you would print some posters and leaflets and go around putting them up on noticeboards and handing them out. It was expensive and time-consuming. Today you can post details of your event with the click of a button and the news will reach thousands of people within seconds.

Thanks to the Digital Revolution, businesses of all sizes can offer products and services to a global market. As individuals, we can gather information and ideas from a whole world of knowledge, via an endless and ever-expanding supply of online resources. We can share our vacation pictures and videos with anyone in the world who wants to see them seconds after we've photographed and filmed them. And we can access, search, and keep vast quantities of information without needing any physical space in which to store them.

IN 2014 IT WAS ESTIMATED THAT THE WORLD'S CAPACITY TO STORE INFORMATION HAD REACHED THE EQUIVALENT OF 4,500 PILES OF PRINTED BOOKS REACHING FROM THE EARTH TO THE SUN!

And how much do we have to pay for all this? Almost nothing.

You might think you had to pay a lot for your smartphone or your TV, but in the late 1970s the price of a color TV was $350 to $500. Today you can buy an UltraHD TV for little more and the quality is light years ahead. Back then, the idea of a phone you could leave the house with, let alone use as a TV, camera, calculator, map, music player, and just about every other function of life, was the stuff of science fiction.

Digital technology isn't just cheap to buy, it opens up a world of better value too. Online shopping and auction platforms enable us to buy goods at very low prices and have them delivered the next day (or even the same day). The level of customer service is incredible. As consumers, who in their right mind would want to roll back the benefits of speed, scale and affordability that digital technology has brought us? But it's not just in the field of consumption that digital technology has brought these incredible benefits.

Information is more freely available today than it has ever been. Doctors and scientists can quickly share knowledge that leads to new cures and inventions that benefit everyone on the planet.

People are able to communicate with one another across political and geographical boundaries. As a result, we can see what's going on in other parts of the world via numerous sources, so we can draw our own conclusions, rather than relying on the broadcast news services, who might have their own agenda.

We can search the Internet for advice on everything from building a house to diagnosing an illness. The Internet has brought out the human propensity for sharing in a way that was hard to imagine 20 years ago. It has also given us immersive experiences that make our choices so much better informed. Planning a vacation? Don't book anything until you've scanned all the options and taken a good look around on their website… or taken a virtual walk around the neighborhood using Google Street View.

Most of these technologies have evolved in the last couple of decades—an era that's become known as the Information Age. Information Technology (IT) has created new career opportunities that simply didn't exist a quarter of a century ago. Data scientists, digital marketers, data security officers, social media managers, app developers, UX designers…

The Digital Revolution has changed the world forever and, for the most part, it's changed it for the better.

DESIGNED FOR LIFE

UX, in case you're wondering, stands for user experience—a term that lies at the heart of the Information Age. User experience is the way any piece of digital technology makes you feel as you use it. You will have noticed that the design of apps and websites has changed a lot and continues to change, becoming more user-friendly all the time.

In the early days of the Internet, websites were often aesthetically unpleasant, awkward to navigate through, and hard to read. Consequently, visitors didn't spend long looking at them. Website developers noticed this and started designing and writing their sites to be much more user-friendly, learning all the time about the habits of their users and adapting their sites to suit those habits.

Another incredible benefit of digital technology is that it gathers its own feedback. Because every action we take on an app, website or game is recorded digitally by the product itself, digital delivers a detailed level of self-analysis, which enables developers to constantly refine their products for a better user experience.

The good news for us as consumers is that the experience of using these products improves all the time. As if by magic we find them giving us what we want, when we want it, and in a way that is increasingly easy to use.

The level of convenience is constantly on the rise.

THIS IS JUST THE BEGINNING

Since the start of the Information Age, digital technology has come a long way in a very short space of time.

But there is much more to come. The Digital Revolution may have given us new tools, but we are still the ones using them. Soon all that could change.

If convenience is your motive, you'll be thrilled at the prospect of new technologies like artificial intelligence, machine learning, 3D printing, the Internet of Things, and driverless cars.

We already live in an age where you don't have to leave your sofa to do the shopping. You can go through the whole supermarket online and have your groceries delivered the same day. But you still have to do the looking and buying online.

The Internet of Things could change all that. Connected TVs, music systems, digital assistants, and security systems are already commonplace. Connected heating, lighting, and electrical systems for the home are selling fast. And soon other domestic appliances, like your fridge, your dishwasher, and your washing machine, will be connected too.

What this means is that they will be able to "talk" to each other and share intelligence without you having to take part. A favorite example is the fridge that places your grocery order for you, because it knows what you usually buy and what's running low.

This sort of automation will develop most rapidly in work applications, where the accuracy and efficiency of machine learning will reap very valuable benefits for businesses. Robots have been building cars for decades, but breakthroughs in digital technology are paving the way for machines to take on manual jobs like construction on a much wider scale.

This will, in theory, leave humans to focus on the more creative jobs and to provide the unpredictable element that is so important for establishing commercial advantage.

The power and accuracy of artificial intelligence will enable us to refine all our commercial ventures to provide customers with exactly what they want, right down to the point of personalizing the goods we buy. While traditional manufacturing methods make personalization hugely inefficient and expensive, new ways of manufacturing, specifically additive manufacturing (or 3D printing), will enable companies to produce highly personalized products without the additional cost.

This facility for personalization will be a serious benefit in the medical field. Currently, when doctors prescribe pills for their patients they are restricted to preset dosages; with additive manufacturing they will be able to prescribe very precise doses, tailored to each patient's specific needs, and delivered to a 3D printer conveniently close to the patient's home.

3D printing is also breaking new ground in the manufacture of prosthetics and could even be used for manufacturing replacement organs, such is the level of detail that's achievable from this technology. This is also being explored in the manufacture of mechanical components, such as jet engine parts, which could result in much faster, quieter, greener, and safer modes of transport.

CONNECTED CARS AND CITIES

Do you ever dream of a world without traffic? Or road accidents? As cars connect to the Internet and start "talking" to each other, the flow of traffic will be controlled by clever digital computations to avoid the stops and starts that snarl up our roads today, and get us safely to our destination.

We won't just be alerted when the car needs servicing, we'll be shown a choice of garages and a comparison of costs and customer reviews, so we don't have to do the research ourselves. And soon, we won't even have to drive the car—we can sit back, relax and watch a movie while the car finds its way from A to B, "talking" all the way to other cars, traffic lights, and satellites to find the quickest, most economical route.

In our cities, the ability for communication between different devices, such as mobile phones, CCTV cameras, and other sensors, will enable greater safety and security, as well as a more cohesive flow of people and vehicles. Artificial intelligence will find cost-effective uses of facilities, such as office space, so buildings don't lie dormant at night but double up as a hotel perhaps or an entertainment venue.

These technologies are within our capabilities now; some are already in operation. Within 20 years they may be so commonplace that they're beginning to look outdated and new technologies that have not been dreamed of yet will be starting to take their place.

Such is the speed of progress within the Digital Revolution.

WHAT'S WRONG WITH THIS PICTURE?

The Digital Revolution has brought many benefits to our lives and will undoubtedly help to usher in further improvements. Yet for many people the prospect of more advances in technology is not a happy one.

While the benefits are plain to see, there is something not quite right—it's an uneasy feeling that this relentless march in the name of progress might just leave you feeling that your quality of life is getting worse.

Technology that we imagined would be laborsaving and allow us more leisure time as a result of greater productivity and making our lives easier has created something completely different. People are working harder, working longer hours, and their lives have become more stressful, not less. Don't misunderstand me, very few of us have to labor 18-hour shifts in the fields or in coal mines anymore, but many of us have become the 21st-century equivalent—tied to our work, rarely, if ever, disconnected from it.

The damaging effects of digital addiction can make you so wary of all digital applications that you become blind to the genuine benefits. In a world that is increasingly run by digital technology, "technofear" can be as debilitating as digital addiction.

The aim of this book is to help you embrace the benefits of IT and avoid the threats. In the next chapter, we will begin to examine what those threats are and where they are coming from.

Chapter 3

THE PROBLEM

IN THIS CHAPTER
•STRAINING YOUR BRAIN •LIMITED ATTENTION SPAN
•LACK OF IMAGINATION •MEMORY LOSS •ANTISOCIAL MEDIA
•AVOIDING REALITY •LOSING SLEEP
•DEATH BY DISTRACTION •GOOD NEWS

Despite the benefits delivered by the Digital Revolution, our mental health is deteriorating. Clearly something is wrong. The more we embrace digital technology, the more we detest it. The easier it becomes to use, the harder we find it to unhook. And the more we speed things up, the less time we seem to have.

If you had to sum up the benefits of the Digital Revolution in two words, you would be hard pressed to find two more fitting words than "speed" and "convenience." Throughout history, the quest to do things faster and more easily has driven the human race to ever more remarkable feats of ingenuity and it continues to do so as you read this.

Yet there is a contradiction: Because if you had to sum up the prevailing condition of human beings since the start of the Digital Revolution, the words "unhurried" and "relaxed" would not apply.

Studies have revealed some very worrying effects from excessive use of technology that should leave you in no doubt that regaining control over your digital use is important for your mental and physical health as well as being important for your relationships.

STRAINING YOUR BRAIN

Research has shown that intensive use of digital devices is having a detrimental effect on our brains. Faculties like memory, attention span, concentration, organization, empathy, and self-control have all been shown to suffer.

The way online content is presented delivers a huge overload of information for us to try to sort out. "Should I click this link or read on?" "Will that link be more useful than this one?" "Have I got time to watch this video?" "Will I be missing out if I don't watch it?"

We like to think we can multitask, but the fact of the matter is that when it comes to asking your brain to concentrate on two things at once, it can't. It is capable of doing things like listening to music and drawing a picture at the same time, but when it comes to cognitive functions like decision-making your brain has to concentrate on one thought at a time.

When you present it with two or more cognitive tasks, it flits between them. This is very tiring and stressful. If you've ever tried writing a letter, say, while listening to an important interview on the radio, you'll know how tense and stressed it makes you feel. Your brain is trying to take in the points made in the interview while also trying to articulate completely unrelated thoughts in writing. You're tearing your brain in two and it hurts! That's not to say that you can't write a letter while listening to talk radio—you'll find that your brain automatically filters out the talk so it can concentrate on the task in hand. Attempt to override that filter and focus on both, and it causes tremendous strain.

We don't just bombard our brains with the multiple decisions presented by looking at a website, say, or social media, we add to the problem by using multiple devices at the same time. For example, you can play an online game while chatting to friends online who are also playing, at the same time as watching a program on your TV and chatting to other friends via social media on your phone.

We take pride in our ability to handle all these devices at once, but what's really happening is that we're handling all of them badly. In the process, we are losing the ability to carry out any cognitive task well. Experiments have shown that people who multitask a lot and think they're good at it actually score very poorly for their ability to switch between tasks, ignore irrelevant information, prioritize, and compartmentalize.

LIMITED ATTENTION SPAN

In the last few decades, scientists have discovered that the brain is very plastic—not made of plastic obviously, but it has plasticity, meaning it can reorganize itself and be molded and remolded by conditioning. Think of it as a network of electrical connections that fire off every time you have a thought, or process information, or pull something out of your memory.

This network changes shape in order to give more capacity where it is needed and less where it is not. If you are fed a lot of information on one particular subject, the network will remold itself to accommodate this information.

So the more we use our brains for a given task, such as playing the piano or remembering phone numbers, the better we will become at those tasks. By the same principle, the less we use our brains for certain tasks, the worse we become at those tasks.

Digital devices require our brains to practice shallow thought, constantly flitting from one quick decision to another. The result is that we are losing our ability to have deep, meaningful, considered thoughts—to focus, concentrate, work out solutions, or maybe to completely lose ourselves in a book, or movie, or in music. Digital life risks us becoming superficial; digital addiction guarantees it.

As intelligent creatures, we have a natural desire for information. Whether that's share prices, football scores, or the latest showbiz gossip, we find the promise of new information hard to resist. When you look at the screen on a digital device you are inundated with such promises. This makes it incredibly hard to concentrate on the subject you came to the page for—you find yourself in a snowstorm of distractions. It's like a classroom of toddlers all poking you for attention.

Every time you give in to a distraction, you condition your brain to give in, so you become less able to not give in. In other words, you become mentally weak. You might not be fully aware of the stress at the time, but it takes its toll. Your brain tires, your focus wavers, you make poor decisions, you become irritable. Over time, your brain's capacity for concentration is diminished and your overall mental ability suffers.

We forget some simple but beautiful pleasures such as losing ourselves completely in a movie.

LACK OF IMAGINATION

Watch an athlete before a race. They will go into a world of their own. They might listen to music, close their eyes, cover their head, look to the skies… anything to help them clear their mind of the distractions around them. What they're trying to do is get into "the zone."

If you're not familiar with "the zone," it is the term used for that mental state when everything just seems to fall effortlessly into place.

Writers have historically referred to it as "the muse"—a divine spirit that endows you with the inspiration to create great work. But you don't have to be an athlete or an artist to experience the zone: It is a state of mental focus and clarity that enables anyone in any walk of life to produce their best work and achieve great things.

Digital devices destroy your ability to get in the zone. The zone requires total focus. Distractions take you out of the zone. In other words, they prevent you from doing your best work and deprive you of your best moments in life.

Smartphones have filled the emptiness in our lives. That might strike you as a good thing. We regard emptiness as a negative. We feel uncomfortable with silence, we become restless with time ticking by unfilled and we go out of our way to avoid boredom. Digital devices enable us to be working at something 24/7, filling our time with something apparently constructive. But how constructive is it really?

That phone call you make from your car, just to eat up the journey time—is it really necessary? Does the person you're calling want to take that call? That manic scrolling through your social media feeds— is it improving your quality of life?

A bit of emptiness is actually very good for you. It leaves room for your imagination to flourish, ideas to hatch, or simply for your brain to rest. Without emptiness, your ability to be creative is severely diminished. Smartphones may be the work of some highly creative minds, but they are destroying creativity in their users.

MEMORY LOSS

What were we talking about again? Oh yes. Digital devices are having a negative impact on memory, both short term and long term. There are two classic situations where the human memory typically struggles:

One is numbers; the other is names. Do you know your own cellphone number? You probably do. But what about your partner's number? Or your kids' numbers? A lot of people don't. Have you ever suffered the embarrassment of being introduced to someone at a party and then going to introduce them to someone else and finding you've forgotten their name already? Within the space of three minutes!

But digital technology has created something that we find even harder to remember than names and numbers: passwords. How many times have you gone to enter your password and found yourself blocked out because you've gotten it wrong somehow? Passwords are notorious for constantly taxing our memories and it's no wonder really —passwords are a combination of names and numbers!

Fifty or so years ago it was calculated that the average person could memorize roughly seven to nine digits, which was fine because phone numbers tended to have seven digits at most. Today, our memory capacity is estimated to be around five digits!

You might not regard this as a problem, since our phones do much of the remembering for us, especially phone numbers. But that's part of the problem: The "muscle" is not being exercised and so it is becoming weaker. At the same time our memories are being constantly bombarded with new information and distractions. Like a waiter trying to carry too many plates, we end up dropping them all.

ANTISOCIAL MEDIA

We seem to have developed a love–hate relationship with social media in the short time that it has been in existence. We complain about the intrusiveness, we worry about the Big Brother effect of giant corporations having access to so much of our personal information, we sneer at the sugary posts other people put up, become infuriated by political views that friends, family, and friends of friends thrust upon us, and feel intimidated, resentful, or envious of the perfect personas some of them present.

But is social media really so bad? The simple answer is yes!

When you sign up for a social media account, you are willingly giving yourself up to be a guinea pig. That's why all these apps are free. They don't make their money from users like you; they make it from Big Data.

Commercial organizations have always tried to learn as much as they can about their potential customers, so they can target their advertising as accurately as possible to your tastes. Big Data harvested from social media has enabled customer profiling to become so accurate that you now receive adverts for products you might want before you've even thought about them! Annoyingly, you also receive adverts for products you've just bought! Not so clever.

More dangerously, in the age of the social media influencers, most commonly via Instagram and YouTube, many people aren't even aware they're being subjected to advertising by the "influencers" who accept payment, freebies, or other gifts from a whole host of High Street brands and global corporations in return for apparently unsolicited endorsements of their products.

But this profiling makes Big Data hugely valuable. And in order to keep harvesting it, the app designers do everything they can to keep you hooked in.

Social media is incredibly addictive. Once you're hooked in, it becomes very hard to switch off for fear that you might miss something that affects your social life. The irony is that, when used indiscriminately and aimlessly, social media is killing social life. Its effect on real-life relationships is destructive. For every second you spend looking at your social media accounts, you are spending a second less paying attention to the people around you.

How do you think that makes them feel? And how do you feel when people do it to you?

The restaurant table scenario is the classic example and it is all too commonplace. It's an everyday occurrence to see people sitting together, all looking at their phones rather than socializing in person with one another.

Why does this matter? Because statistics show that socially related mental health problems, like depression, bullying, loneliness, and suicide, are rising at an alarming rate.

Social media is a breeding ground for anxiety. We all want to be accepted by our society and we try really hard to present a likeable persona on social media. And we are given instant feedback as to just how successful our efforts are by the likes, comments, and follows that are intrinsic to social media apps.

While we constantly compare our social personas with everyone else in our network, we forget that they are carefully manipulating theirs too.

The result is that everyone on social media is living in a fake world of carefully curated unreality, where everyone looks fantastic, has fantastic experiences, and thinks deeply moving thoughts.

Aspiring to this perfect world inevitably causes a sense of inadequacy that can quickly lead to anxiety and depression, a state of affairs which is exacerbated by the appalling impact that unfettered use of social media has on an individual's attention span, ability to relax and unwind, and sense of calm reflection.

This can cause a sense of being made to feel isolated from the group, something that happens regularly among schoolchildren but also among adults on social media. People have always had the capacity to be cruel and scapegoat others, but social media gives us the ability to gang up, ostracize, ridicule, insult, abuse, scandalize, and blackmail, all from the virtual safety of a digital device.

It doesn't have to be like that though. Used proportionately, social media can enhance your life, nurture relationships with friends and family who live in different parts of the world, and be a wholly positive thing.

Social media has become the victim of two paradoxes:

1. It is sold as a social tool, but it can actually make us more antisocial.

2. It is sold as a news-gathering tool, but it actually blurs the line between truth and fiction.

The reapers of Big Data aren't interested in the truth; they're interested in clicks. So those story links you see on social media and other news sites are not designed to inform you about what is going on in the world but to make you click on them. And if that means making up a story that's juicy enough to make you click, that's what they'll do. Being able to sort truth from fiction is becoming increasingly difficult and this is something you need to have good practices for handling if you want to continue using social media once you're free from digital addiction.

Social media has created a false social environment with fake personas, fake opinions, and fake news. It has made us switch off from the real people around us and feel paranoid about the "friends" in our network. The result is anxiety, feelings of inadequacy, isolation, and confusion, and, ultimately, depression.

AVOIDING REALITY

Online gaming designers have done such a good job of replicating the things that give us a genuine sense of accomplishment—skill development, decision-making, construction, leadership, teamwork, speed, efficiency, status, conquering, winning, even to the extent of killing avatars—that an increasing number of young people (predominantly young men) are giving up on living in the real world altogether.

Research in the U.S.A. has found that young, poorly qualified men are spending an increasing amount of time out of work. In 2017, more than one in five men aged 21–30 without a degree reported going the whole year without working at all. In 2000, that figure was less than one in ten.

Researchers concluded that the addictiveness of online gaming was causing people to drop out of the job market.

For young men in their twenties, the idea of playing video games rather than going out to work or going out looking for work may seem very attractive. But the inevitable consequence is a generation of men going through life without any work experience, real-life practical skills, or real-world social skills.

Gaming addiction, known as Gaming Disorder, as recognized by the WHO, is defined as "when an individual plays video/online games to the extent that it significantly impairs personal, family, social, educational, occupational, or other important areas of functioning."

It's not just young men falling prey to this addiction, but young women too, and surprisingly an increasing number of adults in their 30s and 40s are sacrificing career ambition, relationships, family life, and real-life socializing in favor of 20, 30, 40, or even 80 hours of online gaming time each week.

Even addicts at the lower end of the gametime graph, who just about maintain jobs and relationships, cause themselves tremendous harm. There's a huge difference between activities and pastimes such as reading or watching TV—both relatively passive and rather ambient, relaxing pursuits—and 20 hours per week spent fighting to survive in a world full of murderous zombie beings, planning and conducting missions, collecting resources, building fortifications, and constructing weapons and traps to engage in combat with waves of the creatures attempting to destroy everything and everyone. The journey from zero to hero and back again within the game starts all over again once the gamer is dragged from their online world of endless, repeated gratification by their neglected partner or children back into the real world.

That a generation of youngsters is growing up to consider this kind of "reality" as normal is worrying—to think that there are fully grown adults living among us suffering this addiction is truly miserable. If you recognize yourself in this gloomy predicament then congratulations on picking up this book—freedom awaits—and unlike much of the social media and technology referenced so far, the only way of you escaping your gaming addiction is permanent abstinence. Once you realize that the games you've been hooked on are rigged to entrap you forever, constantly redeveloped and redesigned to maintain that objective, then in your heart you know that this one final mission, to escape the clutches of online gaming addiction, and reclaim your "real life," your "real friends," and "real loved ones" is one that you need to complete now.

The great news is that once you commit to escape, and understand how to do it, you'll find it ridiculously easy and actually enjoyable. All you need to do is follow the instructions in this book.

LOSING SLEEP

Digital addiction causes a feeling of excitement and anticipation at the mere thought of looking at your phone. Yet most of the messages you get are disappointing at best. They either offer you nothing or demand something from you, which puts you under pressure and causes stress.

It's like rushing to the front door every time you hear the letterbox and finding nothing but brown envelopes. Junk mail, bills, demands… it's a constant flood of disappointment, taking your excitement levels up and then bringing them crashing down. This is mentally exhausting.

In fact, we should take that analogy further: It's like visiting your doormat every few minutes to check if any mail has arrived, and as you descend deeper and deeper into addiction the volume of junk mail increases in frequency and you become more and more desperate for something stimulating.

It's not just the junk-trawling on social media that robs us of sleep—how often are you drawn into your Insta feed or YouTube, or messaging groups, or online gaming so that you are kept occupied way beyond the time when you might usually want to retire to sleep? You look back on the previous few hours knowing that it wasn't even productive or enjoyable or relaxing… the phrase KILLING TIME could have been invented for the situation. At least if you'd watched a movie, or read, or chatted with your partner, or even made love, you'd look back on those hours as having been relaxing, enjoyable, or fun rather than experiencing that empty, guilty, exhausted, hollowed-out feeling of gloom. As you set your alarm and calculate how little sleep you might enjoy before it wakes you, do you feel full of joy and happiness at an evening well spent? Or do you sigh inwardly knowing that as soon as you drop off, it will feel like a few short moments before you're rudely awakened? That's if you can even get some sleep.

No wonder digital devices are destroying sleep. We live in what we consider to be a stressful world, in which sleep deprivation is an increasingly common problem. But it's no coincidence that the inability to get a good, solid night's sleep has seen such a dramatic rise in the digital age.

Digital addiction destroys sleep due to a number of factors. One is the constant bombardment of messages and alerts. Whether they are stressful messages and alerts or dull ones, it creates a constant, on-guard, braced-for-action demeanor. It eradicates the process of gradually "winding down" during an evening that culminates in a gentle, pleasing, restful descent into sleep. Instead, it becomes the emotional rise and fall of excitement followed by disappointment, and vice versa.

Have you analyzed how bizarrely low your standards of what constitutes "excitement" have become? Someone you don't know, whom you've never met and never will meet, "likes" something you've written or a photo you've posted and it fills you with glee. Yet when the kids rush in from school and want to hug you or your partner wants to snuggle up with you or to just be around you—REAL JOY—you can't wait to get them out of your hair so you can carry on with your online gaming mission or pointless chit-chat.

When we try to justify our excessive tech use, we use excuses like "it relaxes me," "it helps me to chill out," "it helps relieve stress," or "it helps me wind down and get to sleep." In truth it doesn't do any of those things—quite the reverse. It's time to begin seeing your digital addiction as it really is: An endless chain of feeling worse and worse about yourself, drifting further and further away from the things and people that really mean something to you, destroying your ambition, your rest-time, and your sleep, as well as your relationships.

And then there is the blue light.

Have you ever stopped to wonder why you feel tired at night and wide awake on a bright, sunny day? It's not just to do with your body clock. Your brain is designed to react to light and dark detected through your eyes. At the first light of day, it releases stimulating hormones like cortisol and suppresses the sleep hormone melatonin, thus waking you up and making you feel more alert.

The release of melatonin is key to feeling sleepy at night and getting a good night's sleep. But this is where the blue light comes in. The light emitted from screens is a blue light that has the same effect on your brain as daylight—it fools your brain into thinking it's still daytime, and so the release of melatonin is supressed. That's why if you stay up late looking at screens, you are subjecting your brain to endless daytime. And that's mentally and physically exhausting!

Sleep deprivation doesn't just make you feel exhausted, it can have serious mental and physical effects. It blunts your judgment, your mood, your capacity for learning, your concentration, your memory. Bad decision-making, carelessness, accidents, and arguments all become more likely when you're tired.

Sleep deprivation is also linked to chronic physical conditions like obesity, diabetes, and cardiovascular disease. Your digital screens aren't just making you a bit tired, they're gradually destroying you!

DEATH BY DISTRACTION

In 2017 in the U.K. it was reported that somebody was killed every ten days by a driver distracted by their cellphone —a 50 percent increase on the previous year. In the U.S.A., nine people a day are killed as a result of a distracted driver.

Among teens, the number of drink-driving fatalities has decreased, yet the number of driving fatalities overall has not. Something has replaced alcohol as the killer on the road. Distracted driving is hard to prove, which means that the actual figures will be much higher than those reported. Like drink-driving, people who are in the habit of using their cellphone while driving will go on doing it until they are caught or until they crash. If you allow your phone to be a distraction while you're driving, you run an increasing risk of killing somebody or killing yourself every time you get in your car.

New phones now include apps that detect when you're driving and shut down all their notifications. This is a step in the right direction, but it serves to illustrate that digital addicts are considered incapable of self-regulation, even when it comes to a life and death situation like driving. If you need an app to keep you concentrating on the road ahead rather than your smartphone, then you are not in control. Of anything!

GOOD NEWS

The purpose of this chapter is not to fill you with gloom. The good news is that all these harmful effects are reversible, and you will see the improvement as soon as you finish this book and become a Happy Digital User. To reach that goal, it is essential that you recognize and understand how low this condition has dragged you down. Earlier I promised not to bore you or patronize you by telling you what you already know, the harms and dangers that you are suffering as a result of your digital addiction—but hopefully you understand that to truly appreciate the wonder, the joys, and the sheer pleasure of where you are headed, FREEDOM, you need to look at the addiction in its true light.

Perhaps you've already considered the impact that your addiction is having on your partner or your kids. Most likely it's one of the factors that has driven you to reading this book. But deep down inside, perhaps you harbor some resentment. Surely it's their expectations of you, rather than your behavior under the influence of this addiction, that's the real problem? Why can't they be more supportive about what you choose to do with your spare time? If you still feel a little like that, then you're not seeing your existence, your decline, your demeanor, and your lack of vitality through their eyes.

THIRD INSTRUCTION:
SEE YOUR SITUATION FOR WHAT IT REALLY IS

The problem isn't with other people's perception or opinion of the way you live your life—the problem is with the way you live your life.

The purpose of this book is to help you regain control, access all the benefits of digital technology, and avoid the traps. This means changing your mindset.

Smokers who quit with Easyway learn that the only way to stop is to cut out cigarettes, e-cigarettes, and all nicotine products completely. There is no scope for having "just the one" every once in a while, because "just the one" is enough to get you hooked all over again. That's simply the way that drug addiction, all drug addiction, works. Once the smoker understands that smoking does nothing for them whatsoever, it's easy to stick to.

When it comes to digital addiction, the distinction between smart use and dumb use needs to be defined more carefully. We have established that there are many benefits to digital technology and the task is to distinguish between the applications that give you a genuine benefit and those that merely give you the illusion of pleasure or comfort. Let's call them "junk apps." They are the junk food of the digital menu. So how do you tell the real benefit from the illusion?

That is a decision for you to make, and later in the book you will be asked to give this more thought. You will also be given the tools to help make the decision for yourself.

The most important and powerful tool of all is your own mind. The problem is your mind has been taken over by brainwashing. In the next chapter you will learn how that happens and why—it is an extraordinary catalog of tricks and cons. The good news is that while you may feel like a complete slave to your devices and apps now, you still have the power to reclaim control. To achieve that, you need to follow the next instruction.

FOURTH INSTRUCTION:
OPEN YOUR MIND

What does that mean? It means be prepared to accept the possibility that what you're told in this book may be true, even if it seems ridiculous or far-fetched at the time.

If that sounds like a recipe for brainwashing, think again. Brainwashing occurs when you are presented with only one version of the facts—a false version in the case of addiction. What Easyway asks you to do is open your mind to all versions of the facts and then allow your own logic and instincts to distinguish truth from lie. If you persistently attempt to resist or dismiss the information you read here, then you will create a mental block that will stunt your growth. However, far from me asking you to accept without question everything you read, I simply want you to be open to the idea that what you read *might* be true. Can you see how a closed mind is simply incapable of doing that and how that leads to nowhere?

You may already regard yourself as an open-minded person, but, like everybody else, you will have lived your life with your mind largely made up by other people. Let's face it, you can't learn absolutely everything from personal experience. You have to build your knowledge on the things you're told, rather than finding out for yourself.

Here's an example. When you see the sun rise in the morning, you perceive it as a ball of fiery gases burning millions of miles away that has the appearance of rising in the sky because the Earth is turning. This is an established fact, but how do you know that's really the case? Have you been to the sun? Of course not, you've just been presented with some convincing arguments by people who you regard as experts in that field.

But it wasn't very long ago that most people were convinced the sun was actually a god driving a fiery chariot across the sky. That was the explanation put forward by the learned men of the time and it seemed to fit with what they saw.

The point is that people are swayed by the most convincing evidence presented to them and most of what we believe is based on information presented to us from other people. When we are presented with only one version, it leaves us vulnerable to brainwashing.

Here's another example. Take a look at the two tables opposite, one square, one rectangular.

Have a guess at the dimensions of the two different tables. Now take a ruler and measure them.

Incredible, isn't it?

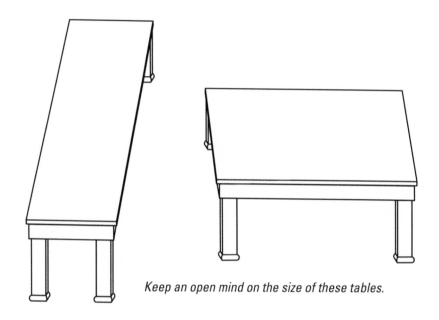

Keep an open mind on the size of these tables.

The two tables are identical, yet you assume they're different sizes because they look different and you were told they were different: one square, one rectangular. When the information you're given tallies with what you see or feel, why would you question it?

This is where an open mind is important. You might have seen through the illusion for yourself had you been presented with a more open question: For example, "Do you think the tables above are different shapes or identical?" This would have set your mind questioning what your eyes were seeing. But when told the two tables are different, you accept it because the illusion makes them look different.

We assume there is one explanation for everything we see or feel, but this is not the case. Addiction thrives on this assumption. It presents us with one explanation and relies on us never to look for another.

Easyway works by presenting you with the alternative explanation. All you have to do is open your mind to the possibility that there is an alternative explanation and then, keeping your mind open, allow logic and your own senses to help you see which explanation is the true one.

There were two explanations for the two tables you saw: One was that they were different, one square, one rectangular; the other was that they were identical but had been drawn cleverly to create the illusion of being different. The latter explanation seemed less plausible at first, but by measuring the tables you now know that to be the true one. And now that you know, you can never be fooled by this illusion again.

This demonstrates how your mind can easily be tricked into accepting false "facts." When you first started getting sucked into tech addiction, you probably believed you were exercising your freedom to choose what you wanted to do and when, but what if you were basing your choice on false information?

As you continue through the book, remember the table illusion and remind yourself to keep an open mind, so that even if something comes up that you find difficult to believe, you will accept the possibility that it is true.

Now, let's take a look at those tricks that got you hooked.

Chapter 4

RATS IN A CAGE

IN THIS CHAPTER
•*NO ACCIDENT* •*BRAINWASHING* •*HOW ADDICTION WORKS*
•*TRICKS OF THE TRADE* •*SOCIAL MEDIA* •*THE FIFTH INSTRUCTION*

The pull to keep checking your phone is like running to the front door every time you hear the letterbox – the result is nearly always rather disappointing. So why do you keep running to the front door?

The addictiveness of digital devices is very strong. Every time your phone emits a noise or a vibration, you instinctively reach to pick it up, just as a laboratory animal can be programed to rush to the food bowl every time it hears a bell ring. Smartphones have turned us into lab rats.

In fact, these instinctive responses have always been there. What has happened is that digital devices have brought them out in ways that we have never seen before.

This is no accident. There is a vast and growing industry that relies on you responding the way you do to certain tech triggers. The more you use junk apps, or use good apps with a junk attitude, the more of your data they can gather, the more products they can push your way, and the more money they can make.

This is a multi-billion-dollar industry and it attracts some of the brightest minds on the planet. They keep themselves amused (and rich) by studying the way our brains work and developing apps and alerts that exploit our temperament, instincts, and way of thinking to trap us in cages. They know more about the way your mind works than you do, so they can control you without you realizing you're being controlled.

Here's an everyday example.

Once upon a time there were TV serials that aired once a week, and so you had to wait a week for the next episode. Now, when one episode finishes, you have to actively make it stop or it will automatically start playing the next episode. The digital designers know that we are fundamentally lazy when it comes to calling a halt to something we're enjoying and they have cashed in on this by spoon-feeding us content in a way that we find irresistible. Thus they can make you consume far more than you planned to when you sat down. It sounds very similar to junk food, doesn't it? It's the equivalent of binge eating, except there is no physical limit to the amount you can consume.

It's perfectly reasonable and normal to watch a YouTube clip that someone sends you because they think you'll be interested in it. But the pressure is to watch it now and give feedback on it straight away… and then to share it. Even then, while you're still mentally digesting the clip that you've watched, another clip starts up automatically—quite often on an unrelated subject, and so the potential to be kept on the YouTube clip treadmill is created. From one shallow distraction to another, to another, we suddenly realize that we've wasted hours without even feeling rested. The changes in mental trajectory, never settling down for more than a few minutes as we bounce between subjects, make us feel like the ball that's smacked in all directions in a pinball machine.

It's not relaxing; it's exhausting.

Most digital users don't realize they're being controlled, but there is a suspicion—especially when your digital use spills over into addiction. You know you're not in control because when you want to stop you find you can't, or it's extremely difficult and requires all your willpower. You're occasionally aware of the discomfort the addiction is causing when you have to keep checking your phone even on short journeys. Whatever happened to the simple process of looking out of the window of a car, or a train, or a bus for a few aimless moments lost in thought, rather than head constantly bowed down at the screen of the slave-master: digital addiction?

You battle against the slavery, but you can't apply that level of willpower forever, so eventually you slip back into feeling controlled. The effect is devastating. The inability to unhook causes feelings that are remarkably similar to those of a drug addict. The sense of powerlessness makes you short-tempered, then depressed, then devious and deceitful. And if anybody tries to take the device away, or lure you from your games console, you become resentful, angry, and aggressive.

You want to resist the urge to look at your screen, but some unseen force overpowers your will and you're faced once again with the realization that you've become a slave. No one wants to feel that way, so you do the only thing you can do when faced with a truth you don't want to accept: You go into denial. You kid yourself that there's nothing abnormal about your digital use and that you could put a stop to it any time you wanted to.

IF YOU CAN STOP BEING A SLAVE TO DIGITAL ANY TIME YOU WANT, WHY DON'T YOU STOP IT NOW?!

After all, it's not making you happy, is it? What's holding you back?

The reason you can't stop when you want to is because you don't really want to stop. That might sound contentious. After all, why would you be reading this book if you didn't want to stop? It's at times like this when you need to remind yourself of the fourth instruction: **KEEP AN OPEN MIND**.

The problem is that, despite what you think about digital addiction and how it's blighting your life, deep down you believe that it's giving you some sort of pleasure or comfort and that if you stopped using it, life wouldn't be quite the same.

You're absolutely right.

> *WHEN YOU QUIT BEING A SLAVE TO DIGITAL ADDICTION, LIFE WON'T BE THE SAME. IT WILL BE BETTER... INFINITELY BETTER*

BRAINWASHING

It's very easy to believe that constant use of digital devices and apps is necessary if you want to keep up with modern life. These devices and the junk apps they give you access to are constantly changing and that puts a pressure on consumers to keep up or crash out of the game. Again, this is no accident. It is a thoroughly researched marketing ploy.

It is almost impossible to live without digital devices these days. There are very few jobs left that don't rely on desktops, laptops, phones, or tablets every day.

On top of that, your phone, Internet, email, and social media have become your connections to the world you want to live in: your friendship groups, your family and shared interest groups. But when they become constant sources of intrusion, demanding attention, sometimes on a second-by-second basis, that's when you know you have a problem.

When you want to kick back and unwind, do you turn everything off and do nothing? Or do you pick up your phone and scroll through your social media and messages?

All addictions are caused by brainwashing. Smokers are brainwashed into believing that sucking foul-tasting smoke into their lungs is stylish, sophisticated, good for their self-confidence, concentration, and relaxation.

As a digital addict, you have been brainwashed into believing you cannot live a meaningful, connected life without your phone, every moment from the time you wake up to the time you lie down to sleep.

The challenge, as with every addiction, is to reverse the brainwashing. That's simple to do once you separate pleasing, sensible, necessary, and beneficial use from permanent, constant connection.

Once you can see this, the desire to repeatedly look at your phone, to crave time for online gaming, and being automatically enslaved to social media, simply fades away. Does it take effort? No more effort than it takes to stop checking the letterbox at the front door every few minutes to see if any letters have arrived.

On the face of it, curing digital addiction is a little more complex than curing, say, smoking or cocaine addiction for two reasons: First, as we have established, there are many benefits to digital technology, so the aim is not to make you shun digital devices completely; second, the brainwashing is even more cunning and relentless than for any other addiction we have ever seen.

This might leave you feeling that you're pursuing a hopeless cause. Rest assured, you are not. The brainwashing may be cunning and relentless, but it's just as easy to reverse it as it is for any other addiction. It is a simple two-step process:

Step 1: Recognize and accept that you've been brainwashed.

Step 2: Follow Easyway to unravel the brainwashing.

HOW ADDICTION WORKS

When you do something that gives you pleasure, such as eating a nice meal, playing a sport, going dancing, or making love, that feeling of pleasure is caused by chemicals released in your brain. The pleasure system is an ingenious part of our biology, designed to encourage us to seek out the things that give us pleasure because these are the things that help us to survive: eating, drinking, staying fit, reproducing…

Because these things are fundamental to our survival, the pleasure chemicals are not only released in response to doing these things (a reward), but also in anticipation of doing them (an arousal). For example, we begin to salivate at the thought of a delicious meal. This reaction is all part of our survival mechanism that compels us toward behaviors that are good for our survival.

Let's call it excitement. It's what drives us to seek out food and water, competition, sexual partners, and all the things we desire.

Addictive behaviors and substances hijack this system by triggering a release of pleasure chemicals. Not because they deliver genuine pleasure—in most addictions, contrary to popular belief, that's simply not the case—but because the ending of the discomfort of withdrawal mimics genuine pleasure.

This programs the brain to get excited at the thought of these behaviors, and so you are compelled to repeat them—never realizing that the so-called pleasure associated with taking the drug is nothing more than a con. It's like wearing tight shoes simply for the relief of taking them off.

Think about your phone now and pay attention to the way your mind and body react to that thought. Your heart rate increases and you become excited at the thought of looking at your screen. This is the impulse that keeps you reaching for your phone, even when you're desperate to quit.

There is an obvious question that arises at this point for any addiction: "If it gives you pleasure, why is it a bad thing?"

The "pleasure" you get is not a genuine pleasure, in the way that quenching your thirst is; it's a relief from discomfort. Addiction works by making you feel uncomfortable (withdrawal) and then partially relieving that feeling (the fix). It is the illusion of pleasure and relief, rather than genuine pleasure and relief, that keeps you coming back for more.

Furthermore, the addictive behaviors that create the illusion of pleasure are not essential to your survival; they are detrimental to your health and happiness. When you satisfy hunger or quench your thirst, your body tells you when you've had enough. With digital addiction, there is no mechanism to tell you when you've had enough. Your eyes might grow tired, your body might ache and creak because you've been online for so long, but still you continue. There's simply no respite. The nature of addiction is to take more and more, and digital devices are designed to make bingeing easy. That's not just a box set marathon, or constant social media engagement, but the way online games are designed to prevent players from quitting.

Like all addicts, when confronted with your behavior by a loved one, you push back, resentful and defensive. In the rare moments when you are honest about your tech use, you blame your binges on a lack of willpower. Bingeing on anything always leaves you feeling bad. Feelings of guilt, weakness, and disgust come with a complete lack of satisfaction. You binge in order to feel satisfied, but, as you will discover later, the nature of addiction is such that the more you take, the further you get from feeling satisfied.

ADDICTION COMPELS YOU TO PURSUE AN EVER-MORE-DISTANT GOAL

It's not a lack of willpower that causes you to binge, it's addiction forced on you by digital designers, whose job it is to understand human psychology and behavior and come up with increasingly clever ways to hook us and keep us hooked. The Internet has given them a vast wealth of data to work with and they use it as their laboratory, constantly trialing ideas and measuring the response.

All digital consumers are part of a constant campaign of controlled experiments, a relentless sequence of A/B testing designed to learn all about our preferences and behaviors. It's all about knowing the customer to the nth degree—the Holy Grail of marketing. They know us better than we know ourselves.

TRICKS OF THE TRADE

Understanding what the digital designers know about you and the way your mind works will help you as you prepare to fight back. It's important to recognize these traits in yourself, so you can begin to understand your own responses and control them.

We hate vacuums

Before we had access to cellphones, significant portions of each day would be spent in a vacuum. Walking down the street, waiting for a bus or sitting on a train, waiting for an appointment, driving to a meeting… If we could, we would pass the time by reading a book, or a newspaper, or listening to music. Today phones fill the vacuum.

We've become less used to being alone with our own thoughts and the less we do it, the worse we get at it. In one experiment carried out in the U.S.A., volunteers actually chose to give themselves electric shocks in preference to spending 15 minutes doing nothing!

Digital designers have made sure your phone always offers something for you to take your mind off the nothingness. If that's as bad as it got, it would be tough enough! But not satisfied with keeping you busy on just one thing, your phone is designed to constantly interrupt whatever it is you are doing on it with a whole range of alerts, notifications, and messages from the variety of apps that are installed on it.

There's nothing unhealthy or worrying about reading your daily newspaper or a book on your phone on your morning commute, but you certainly don't need to be interrupted by alerts reminding you to review your last Uber trip, or telling you your credit card statement is ready, or that something inconsequential has happened in the news, that someone's liked something on your Facebook page, or that your photo app is suggesting that you post a selection of photos from the past few months, and so on.

You really don't need to know any of that stuff, so how does it get to interfere with your reading pleasure? No wonder you're exhausted. This isn't just going on on a daily basis; it's happening hour after hour, day in, day out!

News alerts are particularly annoying. The majority are entirely pointless and uninteresting. I remember the days when a news alert would only occur when a president was shot or some other huge event occurred; these days a politician blowing their nose seems to warrant alerts from all media outlets direct to your screen.

Rather than having moments of reflection, moments when you're doing nothing and allowing idle thoughts or creative ideas to develop, your consciousness is constantly filled to bursting with banal alerts, gifs, memes, and information you simply don't need.

We love new things

Whether it's toys, clothes, or cars, we always get more excited by things that are brand new. How excited were you last time you bought a new phone and unwrapped it? The packaging, design, and even the instructions are carefully designed to stimulate your excitement. And even though current smartphones are incredible machines with capabilities we couldn't have imagined ten years ago, as soon as a new model comes out millions of people will rush out to buy it.

Apps too are designed to give us something new on a regular basis: New content, new friends or followers, new likes and shares... The software has to keep developing too; it can't afford to grow stale because there will always be something new ready to replace it. And we will always be looking for it. We love new stuff and they know it. Yet this leads to constant change for change's sake. Apps and processes that were perfectly intuitive and low maintenance suddenly become confused and complicated with every update that occurs. It takes time and effort to adjust, such are the shifting sands on which our digital life is built.

We love things that work

Give a child a light switch and they will amuse themselves for ages turning it on and off. We are fascinated by things that make things work—cause and effect.

When we press a button and something cool happens, we feel

compelled to press it again. We derive satisfaction from getting a reaction to our actions.

Digital devices are packed with buttons that make things happen. Send buttons, hyperlinks, open buttons, close buttons, swipe left, swipe right… and if they light up or make a whoosh sound too, so much the better. So the digital designers give us buttons that make it really satisfying to click and open and send and share, so that we keep doing just that, over and over again.

We love inconsistency

Guaranteed satisfaction is boring. If you knew that there would be a rewarding notification every time you looked at your phone, you would look at it less. This may sound illogical, but it is a characteristic of human nature that digital designers have tapped into with a vengeance. When the reward is unpredictable, it feels more rewarding.

It's this psychology that lies behind our love of games and sports. When a team wins every week, watching them becomes less exciting— even if it's your team and they play really well. We need an element of failure or jeopardy to make the success feel better.

In the digital world, this is called "gamification"—designing everyday apps to work like a game. You post something on social media and every like you get feels like a goal. Yes! Sometimes you get no likes, sometimes you get lots of likes. The possibility of a reward, rather than the guarantee of one, is what keeps you coming back for more. When you know something good could happen but you don't know when, you are compelled to keep checking. You stay vigilant because you don't want to miss that moment.

All you need is an occasional reward amid a pile of disappointment to keep you on the hook. It's like a game of roulette—you know that ball has to land on your number sooner or later, so you sit through all the losing spins just to be present when that one win comes along.

The digital designers know we love a game so they design their apps to keep us playing.

We hate anxiety

Perhaps surprisingly, given our love of unpredictability, we go out of our way to avoid anxiety. This negative feeling is another aspect of our survival mechanism. It drives the search for food and the flight from danger. But when anxiety is left unresolved, it leaves us feeling highly stressed.

So guess what, the digital designers deliberately trigger your anxiety by providing new information and emotional challenges every time you light up your phone. The anxiety you feel has its own digital-age name: FOMO—fear of missing out. Every time you put your phone down, the FOMO kicks in.

FOMO has always been part of the human psyche. How many times have you stood in a line and looked at the next line, feeling anxious that it's moving faster and you've chosen the wrong line? Or you hear there's a party taking place, but you're not invited? But FOMO is more prevalent now because of digital communications, constantly informing us of things going on that we are not part of.

FOMO makes us cling to our phones, like an umbilical cord connecting us to the lifeblood of social inclusion.

We crave popularity

Closely connected to our hatred of missing out is our need to feel loved. We are social animals and we crave popularity. Digital designers give us that affirmation in the form of likes, followers and all the other elements, but compared to real-life interactions, this is a superficial and impersonal form of expression. It's also highly public, so the jeopardy of being unpopular is extreme and the anxiety attached to it is very high.

Just as a like or follow can make you feel loved, an absence of likes or follows can make you feel utterly unloved. Digital designers are fully aware of this insecurity and they play on it mercilessly, stoking our anxiety to keep us checking our score.

We are encouraged to ask for love and to judge one another, neither of which are good for social confidence. We are compelled to massage our profiles to look as loveable as possible, thus creating a false impression, which is echoed by every other user. So while your social networks may look like a perfect world, they are a false reality in which every single profile is fake. Nevertheless, the appearance of so much perfection makes the real you feel inadequate and leads to depression and low self-esteem.

Does this bother the digital designers? Not at all, because they know what you do when you're feeling low: You reach for your phone.

We love personal service

Personalization is all the rage. Whether it's online shopping or just setting your own home screen, the ability to choose something unique to you is compelling. We all want our own personality to shine through and we want that to be recognized. Personalization provides us with the recognition we need that in turn bonds us to our devices and makes us rely on them more.

But take a closer look at the settings you can personalize. You will find that the digital designers have given you lots of control over the apps that make you want to spend more time on your phone and very little control over those that are more practical tools.

Bear this in mind, because it will help you when you come to sifting the apps and alerts you want to keep from the ones you can do away with.

SOCIAL MEDIA

Your digital devices are packed full of these addictive tricks and nowhere more so than on social media. Watch out for them: the affirmations, the gamification elements, the endless lists of potential contacts... all designed to keep you hooked for ever more.

The social media model is designed to attract as many users as possible and to encourage them to share information with an incredible disregard for personal privacy. You're fooled into thinking you're getting something valuable for nothing, but you're not the customer; you're just livestock, kept fat on a diet of addictive content.

Think of other advertising media, such as billboards, TV, or magazines. It's easy to turn away from these things, but social media keeps you hooked, every day, every hour. It's an advertiser's dream and they pay high prices for access to your data and your newsfeed.

Social media and messaging apps have done something drastic to human behavior. They make makes people do things they wouldn't dream of doing in other circumstances: unwittingly sharing fake news stories from racist organizations on social media and sending naked pictures of themselves on WhatsApp or Snapchat, making bitchy comments on Twitter, stirring up hatred... It is a measure of the detachment from reality that people feel they can do these things and get away with them. Of course, no one gets away with anything on social media because it is there for all to see. And of course there is the lowest of the low, the tragic, cowardly, anonymous social media accounts that revel in cruelty and controversy. The characters behind those are often the saddest cases of all. So infatuated and seduced are they by the need and desire to "get a reaction," they stalk social media trolling for amusement. How empty and shallow must a life be in order to consider that "entertainment?"

Be aware of the tricks that tech designers play, beware of the traps they set, and it will help you as you take control of your digital use. Remember, the goal is for you to use technology because, and when, you want to, not because *they* want you to.

Understanding how you have been tricked into digital addiction is key to your escape. The important fact is that your bingeing and inability to unhook from your devices is not down to you being weak-willed; it is the result of a concerted campaign to hook you. This is good news; it means that you have the power to fight back. All you have to do is follow Easyway.

So if you've been feeling a sense of doom and gloom about your digital addiction, you can now put that aside and replace it with a positive approach to making your escape. You are about to free yourself from a tyrant that has been threatening your happiness and health. The simplest but sweetest of pleasures awaits you—pleasures you'd forgotten even existed: For example, curling up with your partner on the sofa and watching a movie while your phone sits ignored on the kitchen counter. You have everything to gain and nothing to lose. This is an exciting moment. Embrace it and prepare yourself for freedom!

FIFTH INSTRUCTION:
BEGIN WITH A FEELING OF ELATION

Chapter 5

FIRST STEPS TO FREEDOM

IN THIS CHAPTER

• *NEW FRAME OF MIND* • *GETTING FREE AND STAYING FREE*
• *WHY ME?* • *ACCEPTING YOUR CONDITION* • *FIRST THINGS FIRST*

The brainwashing that keeps you hooked to your devices creates some very convincing illusions. The truth is that nothing bad will happen if you stop. Escaping the trap is easy and you have nothing to fear. This is the truth: Now all we need to do is make sure you understand it.

Addiction works in such a way that it makes you feel the cure is beyond your power. You can be desperate to quit but still find it impossible to resist the temptation. You remain imprisoned, feeling helpless, hoping for a miracle.

For some people who quit with Easyway, it does feel like a miracle—a magic formula that makes something they've always found baffling suddenly seem obvious. Please don't be misled by anything you might have heard about Easyway:

1. It is not a secret.

2. There is no magic… it just feels like there is.

Easyway works by first opening your eyes to the nature of your problem and then using undisputable logic to help you see the way forward. It strips away the illusions and replaces them with rational thought and true understanding. With your mindset changed, it becomes easy to remove your desire to mindlessly seek comfort or pleasure in digital devices.

The key to unlock your prison is the set of instructions you receive throughout this book and it must be used like the combination lock of a safe. Each instruction must be understood and applied in order for the combination to work. You already have the first five instructions and your escape plan is under way, but please be patient. The key to your escape does not lie in the final chapter or the first chapter, or any chapter alone.

NEW FRAME OF MIND

In order to remove the underlying desire that compels you to keep checking your phone or logging into your online game, you need to change your frame of mind. The first step is to identify the flaw in your current frame of mind that makes you a digital addict, remove that from your way of thinking and let logic and reason undo the brainwashing.

FALLING BACK IN

An addict is caught in a trap like a cage sunk into the ground. Together with Easyway, you have the two essentials that will set you free: You have the desire to get out and Easyway has the key that will make that possible. All you have to do is follow the instructions.

> However, once you are released there is a further danger: The trap still exists and you have to insure that you don't fall back into it again.

GETTING FREE AND STAYING FREE

All addicts are notorious for stopping and starting again. They make a big effort to quit or cut down and then, when they feel they've regained some degree of control, they reward themselves with a little lapse.

"Just the one, what's the harm?"

The harm is that "just the one" is all it takes to trigger a binge and push you back into the trap. So getting you out of the trap is not enough; we need to insure you stay out permanently.

We can do this by making sure you understand the nature of the trap you're in. It may feel uncomfortable for you to admit you're in a trap. We all like to think we're in control. But if you really were in control, surely you wouldn't feel the need to read this book. You would have your digital use exactly where you want it and you would be happy and relaxed.

It's the refusal to admit loss of control that makes addicts blame themselves and punish themselves by staying in the trap. When you accept that you are not in control and that you reach for your phone or games console because you are addicted, you can take your self-judgment out of the picture and concentrate on curing the addiction.

When you allow the brainwashing to go unchallenged—as most addicts do because they're not aware they've been brainwashed—it warps your perception of how things really are. To truly escape the slavery of being a digital addict, you have to recognize the trap you're in and question the brainwashing.

Unlike the cage in the ground, the addiction trap is not a physical trap but a psychological one. In other words, it exists entirely in your mind. It is an illusion conjured up by brainwashing.

THE TRAP IS VERY EASY TO FALL INTO... AND JUST AS EASY TO ESCAPE

Think back to the table illusion in Chapter 3. All it takes is one false piece of information to make you believe that what you're looking at is two different-shaped, different-sized tables. When it comes to using junk apps, or using apps in a junky way, you have been brainwashed with similar false impressions that have created the illusion that you get some sort of pleasure or comfort from it. Therefore, you believe that going without it will mean missing out, falling behind, and feeling isolated.

It's a confidence trick—and once you see through a confidence trick, you never fall for it again. Turn back to those tables with the knowledge you now have. Can you convince yourself that they're not identical?

WHY ME?

Why does the brainwashing trap some people and not others? Millions of people seem to be Happy Digital Users, able to live perfectly normal lives without ever becoming digital addicts, even though they too have been subjected to the brainwashing.

There are several explanations for this. One is that digital addicts work very hard to conceal their addiction, so there's a good chance that the people who seem to be free from the slavery of digital addiction are actually imprisoned just like you are.

Another theory is that some people are born with an addictive personality. In other words, something in their mentality makes them more vulnerable to falling into the trap. The addictive personality theory is frequently cited, as if it were a proven fact, but it is nothing more than a theory and, as you will discover later in the book, there is plenty of ammunition with which to shoot it down. More importantly, it is of no help whatsoever to addicts trying to escape the trap.

The fact is that anybody can fall into the trap and anybody can escape. It's a simple question of understanding what it is that makes you fall in and what it is that keeps you there.

No one forces you to look at your phone every two minutes. You choose to do so yourself. No one holds a gun to your head to make you waste hours of your life engrossed in online gaming rather than spending time with your partner or kids. You choose to do so yourself. The fact that part of your brain wishes you didn't, or can't understand why you do, doesn't change the situation. You do it because you have a desire to do so.

Desire is what makes you feel deprived when you force yourself to go without. Desire is what makes you feel agitated when you see other people using digital devices and you can't. Desire is what drags you back into the trap just when you think you've escaped.

IN ORDER TO STAY OUT OF THE TRAP PERMANENTLY, YOU NEED TO REMOVE THE DESIRE

The only difference between digital addicts and genuine Happy Digital Users is that the latter don't have the same desire to use junk apps or overuse good apps. That is not to say they are immune to the brainwashing. Somewhere in their minds they too will believe that there is some pleasure or comfort to be gained from them. And there may come a time in their lives when they are feeling low and need something to pick them up and they will fall into the same trap as you.

For now, though, when they weigh up the pros and cons of technology, they conclude that it is a tool that they are happy to use when they have a genuine use or need for it, and put it down when they don't.

Some people are averse to social media, some feel they don't have time for it, some worry about their privacy, and some just aren't interested.

Social media isn't all bad. Why shouldn't it be a great way to keep in touch with friends and family abroad, or to see the occasional photo of their kids, or their dog, or their vacation? Why shouldn't your kids' grandparents see occasional posts from your vacation or day out? There are all kinds of positive applications of social media; getting the balance right is essential.

Less is best. You don't need dozens of Facebook friends. Keeping it light, keeping it to very close friends and family, is all that's required.

Before you despair that you are in no position to apply that sort of control or reasoning, here's some good news for you. You don't need to.

EASYWAY DOES NOT REQUIRE THE POWER OF REASON TO OUTWEIGH TEMPTATION; IT REMOVES TEMPTATION ALTOGETHER

Just as somebody who has never seen the tables illusion is susceptible to being fooled by it when they do see it, people who have never fallen into the trap are still susceptible to the illusion that there is some pleasure or comfort to be derived from digital devices. There is no guarantee that at some point in the future they won't succumb to the brainwashing and fall into the trap themselves.

But when you've been in the trap and then removed the brainwashing, you are in a stronger position than someone who has never been trapped: You are no longer susceptible to the illusions. You *know* that there is no pleasure or comfort in digital addiction. Therefore, your desire is removed for good.

The only relevant difference between you and a Happy Digital User who doesn't binge on social media, gaming, or messenger groups is that they don't have the desire to do so. Neither did you until you fell into the trap.

THE ADDICTION FUELS THE DESIRE

Remember, addicts seek comfort in the very thing that's causing them misery. They just can't see the connection. This is the trap you are in. It is a vicious circle, but you can break it by opening your mind and unraveling the brainwashing.

Thanks to Easyway, there are large numbers of ex-addicts who once thought they could never get free from the trap they were in but have now escaped and have no desire to fall back in.

Soon you will join them.

ACCEPTING YOUR CONDITION

Let's explore the comparison between digital addiction and drug addiction. Here's a description of one of the world's most common addictions.

• Known for its harmful effects on the human body.

• Usually hooks its victims immediately and in many cases they remain hooked for life.

- Dealers hook users and keep them hooked with cheap deals and cunning tricks.

- The more miserable it makes you, the greater your feeling of dependency becomes.

- Symptoms include sluggishness, depression, stress, anxiety, anger, self-harming, shame, guilt, deceit, and isolation.

- Benefits: none.

This reads like a description of a Class A drug. In fact, it's a description of digital addiction. It could be a description of any addiction. You might assume it's a description of heroin or some other hard drug because you automatically associate those drugs with their harmful effects and see their addiction for the vile, controlling condition that it is.

You don't see digital addiction in quite the same light, despite knowing very well that all the symptoms described apply. You've probably joked about the inability to unhook from your phone; you would never joke about the inability to unhook from heroin. But you need to see the two addictions in the same light.

You need to understand that the trap you're in as a digital addict is the same as that of a heroin addict. It is a mental trap, not a physical one, and it is created by brainwashing. In order to escape you need to change your frame of mind. But first it is essential to realize that you are in the trap.

It's easy to recognize the heroin trap. The media portrayal of heroin is quite clear: ADDICTION! SLAVERY! POVERTY! MISERY! SICKNESS! DEGRADATION! DEATH! But the media portrayal of digital addicts is completely different. Happy, cool people, smiling and laughing and looking in control, showing no signs of stress, anxiety, or depression, just having fun and doing well.

The message is clear: "You need tech in your life and the more you use, the better you'll get on."

As you read through this book, these illusions will unravel in your brain so that, instead of seeing your digital devices and apps as a pleasure or comfort, you start to see the true picture: Technology isn't something to thrill you or something that you need constant connection to; it's just a suite of tools that you can pick up and put down at will.

By the time you finish the book, your frame of mind will be complete, such that whenever you think about your old use of tech, junk apps, gaming, and social media, instead of feeling deprived because you can no longer do so, you will feel overjoyed because you no longer have to.

FIRST THINGS FIRST

It's time to decide exactly what sort of digital user you want to be. We have established that abstaining from digital devices altogether is impractical and not beneficial. Now you need to establish your goal.

What does a Happy Digital User look like to you? Think about how you want to feel and write it down. For example, "I want to feel less badgered by my phone." This needs to come from you. Put your phone aside, close your eyes, and notice the peace. Imagine yourself as a Happy Digital User, able to put your phone down and only look at it when you really need to. Think of it like a screwdriver—just a tool designed for certain jobs. There's nothing exciting about it, but it does have its uses. So what are they?

The aim is not total abstention but abstention from junk apps, and junk use, so let's define which of the apps you use are practical and useful and which are addictive and harmful junk. Start a list. Look through the apps on your phone and sort them into two groups: Useful and Junk.

Useful apps are things such as maps, music, banking, weather, transport and shopping, as well as simple communication tools such as Skype, Facetime, messaging, and the phone itself. As you come to each app, take notice of how they make you feel.

The real junk will stick out like a sore thumb. As for games, whether it's something like Candy Crush or whatever the latest trending game might be, or whether it's full-on online gaming via the latest game platform, aren't these just pointless pursuits that waste your spare time and energy, providing you with very little, if anything, in return?

Oh, you might chit-chat with a variety of friends online while you're gaming, but at whose expense? Your partner's, your kids'? If you live alone and you're genuinely happy to do so, that's fine… but is it possible that the reason you live alone is that there's no room left in your life for anything, or anyone else, other than your online gaming?

No doubt you're nothing like a hermit—locked inside seven days a week. But if you're spending a disproportionate amount of time on your own, alone except for people you "meet up with" online, then you know you've lost control.

You might console yourself with romantic notions of being a Neo-like character in *The Matrix*, not realizing that the game is using you rather than the other way around. You have your very own blue pill, red pill moment on the horizon, except it features escape from a phoney, fake, pointless world back to beautiful reality.

It's not just junk apps that you need to list—it's "groups" that you can easily do without. You might have once thought it was nice to be in a variety of Facebook groups relating to subjects or sports, or sports teams, or hobbies, that you enjoy—but is it really? Don't they all just add to the onslaught of pointless information that you're bombarded with? List as many as you can to drop out of.

The same goes for WhatsApp or Snapchat groups with friends. Do you really need to be in so many? Of course not. Don't worry about anyone taking offence at you leaving a group; all you need to explain is that you're trying to trim down the number of groups you're in and that you hope they'll always shoot you a message if they want to catch up sometime. The people who count will be fine and will stay in touch; the ones who don't will disappear from your life without the slightest sense of loss on your part.

As you continue through the book, keep building your list. Once you have drawn the line between the useful and junk applications of all your digital devices, the desirable and disposable groups, you will have overcome the problem of defining exactly what it is you're trying to quit. You will have drawn the line between smart use and dumb use. And you will be one step away from becoming a Happy Digital User.

Right now, the thought of doing away with all those apps and games that you "enjoy" may be daunting. Rest assured, it is well within your grasp. It only feels daunting because you have been hooked into these apps and brainwashed into believing that you need them in your life. In order to unravel the brainwashing and escape, you need to have a clear understanding of the prison you've been locked in.

It's time to take a closer look at the addiction trap.

Chapter 6

THE TRAP

IN THIS CHAPTER
• *ALL ADDICTS ARE THE SAME* • *CHANGING YOUR MINDSET*
• *FILLING THE VOID* • *INGENIOUS BUT SIMPLE*
• *AN INEVITABLE DECLINE* • *SO WHAT'S HOLDING YOU BACK?*

Addiction is like being held in bonds, whereby the more you struggle to break free, the tighter the bonds become. In order to escape you need to understand how the trap works and ease your struggle. When you can achieve that, escape becomes easy.

Addiction is a more commonly used term than it was 50 years ago, but the popular perception of an addict has not really moved on. We may use the term lightly to describe what we perceive to be very "moreish" behaviors, for example chocaholic, shopaholic, workaholic, but in all seriousness we tend not to regard people with these obsessions as true addicts.

A true addict is one of those poor souls whose life is going down the drain due to heroin, alcohol, or some other drug, right? You know that excessive digital use isn't healthy, but you wouldn't put yourself in the same category as them, would you?

Maybe you would. Maybe things have got so bad that you do see your condition as being as desperate as theirs. It's fair to say that most digital addicts do not.

ALL ADDICTS ARE THE SAME

You may not be in the same physical state as a chronic heroin addict, but you are caught in the same trap and it induces the same patterns of behavior: Isolation, frustration, irritation, denial, self-loathing, deceit. And if you don't get out now, you could very well find yourself falling in as deeply as they have.

Fifty years ago, the majority of the adult population took the same ambivalent view of smoking. Most people smoked and, even though there was growing evidence that smoking caused cancer, they regarded smoking as much safer than heroin. We know now that smoking is a far bigger killer than heroin.

That's because far more people smoke than take heroin, you might argue. Sure, but why do people do either when they know it could kill them? Is it because of the incredible pleasure or comfort they get from it? Or is it because they're addicted to the drug?

> *WHETHER OR NOT YOU CHOOSE TO SEE YOUR PROBLEM AS ADDICTION, IT DOESN'T ALTER THE FACT OF THE MATTER, WHICH IS THAT SEEKING PLEASURE OR COMFORT FROM DIGITAL DEVICES TRAPS YOU IN A CYCLE OF CRAVING, DISSATISFACTION, AND MISERY IN JUST THE SAME WAY AS DRUGS DO*

Back when smoking was at its peak, we didn't know how nicotine and other drugs affected the brain. Since then we have learned a great deal about the function of the brain known as the "reward pathways."

By making us feel good when we do something good, the reward pathways encourage us to keep doing things that are good for us. But drugs like heroin, nicotine, and alcohol hijack the reward pathways and deliver a dose of pleasure chemicals. Not because they provide any genuine pleasure but simply because the drug seems to relieve the discomfort of withdrawal. This has the effect of giving us some kind of boost or high, followed by a big low. That's the difference between genuine pleasures and the false pleasures we get from drugs: Genuine pleasures give a more lasting high and don't leave you feeling low.

Now research has found that it's not just drugs that hijack the reward pathways—nonsubstance-related addictions like digital addiction and gambling affect the brain in a very similar way.

When the perceived high wears off, you are left with a restless, empty feeling. In addiction terms, we call this withdrawal. Your brain remembers that looking at your phone resulted in a pleasure reward, so it begins to crave the phone again as the withdrawal feeling kicks in. If you look at your phone during this withdrawal period, the restlessness is partially relieved, and so you are fooled into thinking the device has given you pleasure or comfort.

In fact, all it has done is partially remove the unpleasant low caused by looking at the device in the first place. Happy Digital Users don't suffer this low. In other words, the only reason you keep looking at your phone is to feel like someone who hardly looks at their phone at all! That might sound completely mad, but it just happens to be true.

CHANGING YOUR MINDSET

If the only reason you look at your phone is to relieve the low caused by the feeling of withdrawal from the last time you looked at it, it seems reasonable to assume that all you need to do to get free is to stop looking at your phone. Put up with the withdrawal feeling for the short time it takes for the habit to wear off and the craving will stop.

But we all know that this doesn't work. If it did, you would have quit by now without needing any help from this book. The fly in the ointment is that digital addiction is not just a habit, it's a state of mind that needs to be properly unraveled before you can be free of it.

We've known for a long time that people can be brainwashed by bombarding them with propaganda. Evil dictators have used this technique with devastating effect throughout history. It's only recently, however, that we have begun to understand why we are susceptible to brainwashing. As explained in Chapter 3, the brain is molded and remolded by the way it is asked to perform. This is how people become experts in specific fields; it's also how we become addicted.

The plasticity of the brain determines not only knowledge but beliefs and attitude too. We call it mindset. The brainwashing you are bombarded with about digital apps is designed to give you the mindset that those apps are a treat, a pleasure, a comfort.

OUR TASK IS TO CHANGE THAT MINDSET

Understanding this aspect of addiction means we can apply Easyway not only to smoking, where it began with such success, but to other recognized addictions, such as alcohol addiction and heroin addiction, and also to addictions that don't involve drugs, such as gambling and digital addiction.

You probably know a smoker who has quit but still craves cigarettes weeks, months, even years later. Some poor souls do so for the rest of their lives. This is because they have quit through sheer willpower alone, they haven't changed their mindset, so they continue to live with the illusion that smoking gave them a pleasure or comfort and forcing themselves to go without makes them feel deprived.

With all addictions, including digital addiction, you are brainwashed into believing that your little fix is a source of pleasure or comfort, and those moments of partial relief from withdrawal reinforce the illusion. Thus you are deluded into thinking that happiness lies in the very thing that's causing you misery.

Digital addicts are lured into the trap by brainwashing. Once you're in, it is the illusion of pleasure that keeps you there.

FILLING A VOID

It's helpful at this stage to understand exactly why we are drawn to things that we know to be "bad for us." It seems to go against the survival instinct that compels us to do things that are good for us. In fact, that's exactly what it is.

When you're brainwashed, as we all are by the digitech industry, your intellect overrides your instincts, convincing you that digital devices are a source of pleasure or comfort *despite* the fact that they can have a detrimental effect on your mental and physical health.

In fact, the very fact that digital overloading is "bad" is part of its appeal. We like the idea of being a bit rebellious. It suggests personality, wit, non-conformity, individuality... We want to make our own decisions. It's an urge that is caused by an emptiness that opens up during our youth, starting from birth. This emptiness, or "void", affects all of us to different degrees.

Being born is a shock that leaves us desperately seeking security. We cry for our mothers and they protect us. Our neediness and vulnerability continue through childhood, when we're cocooned from the harsh realities of life in a world of make-believe. But before long we discover that Santa Claus and fairies don't exist.

At the same time we're forced from the safety of home, to school and a new set of fears and insecurities. We start to look more critically at our parents and over time it begins to dawn on us that they are not the unshakeable pillars of strength that we had always thought them to be. They have weaknesses, frailties, and fears, just as we do.

The disillusionment opens up the void in our minds—the feeling of emptiness and need intensifies in adolescence. We fill the void with idols: pop stars, movie stars, TV celebrities, sports players. We create our own fantasies. We idolize mere mortals and try to absorb some of their reflected glory. Instead of becoming complete, strong, secure, and unique individuals in our own right, we become followers, impressionable fans, leaving ourselves wide open to suggestion.

In the face of all this bewilderment and instability, we shut down to the guidance of our parents and teachers and look for a little boost now and then. If that boost comes from a source that our parents disapprove of, so much the better. We've been brainwashed into believing that eating junk food will give us comfort, that cigarettes and alcohol make us feel relaxed and happy, that gaming gives us a thrill, that constant digital use keeps us in the know and in the flow.

When we use these things, we do experience a little sense of rebellion. So we naturally associate these things with relief from the void. From then on, whenever we feel the void, we seek comfort in these apparent distractions. But all these addictions make the void worse.

FILLING THE VOID

Experiments have been carried out that demonstrate just how uncomfortable we are when we have nothing to occupy our minds. In one, people were placed in a room with a button and told that if they pressed the button they would get an electric shock —nothing too severe but painful nonetheless. They were given a demonstration before they went in, so they knew the level of pain.

Nevertheless, when they found themselves in a room with nothing to do, some of the participants started playing with the button and giving themselves shocks. The pain of the shock button was preferable to the discomfort of doing nothing!

INGENIOUS BUT SIMPLE

The trap you are in can be compared to a pitcher plant, that carnivorous, jug-shaped plant, which catches flies with an ingenious and cruel confidence trick. The fly lands on the lip of the plant, attracted by

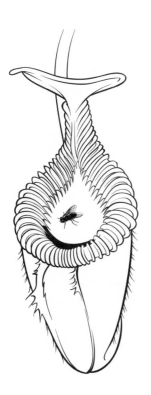

the sweet smell of nectar. As it begins to feed, it doesn't realize that it is being lured further into the plant. The nectar tastes like the best thing in the world, but it is the very thing that is luring the fly to its death.

Before you became hooked on digital devices, you were aware of at least some of the negatives. It's common knowledge that excessive screen use is bad for the eyes; it's antisocial, tiring, and creates a sedentary lifestyle. Also, social media can create a sense of inadequacy, envy, FOMO, paranoia, and anxiety. Yet you also knew that millions of people use digital devices every day without apparently suffering any health problems.

The pitcher plant catches flies ingeniously.

We all like to think we can cope. The human body and mind are incredibly resilient, and anyway, no one sets out to become addicted. So we shelve our concerns about the harmful effects of digital devices, tell ourselves we're in control and we can handle it, and focus on that wonderful pleasure or comfort.

At the same time, the digitech industry bombards us with enticements to increase our use of digital devices.

"Start…"

"Think different"

"Be what's next"

"Greatness awaits"

"Jump in"

"Be connected"

"Join the conversation"

These are all catch lines that have been used by some of the biggest digital companies in the world. They are accompanied by images of smiling, happy, cool people, luring you into the trap. And so you "jump in."

You start devouring digital apps and, guess what, nothing bad happens. So you do it again. But without realizing, you start to increase the quantity and frequency of your consumption.

Just as drug addicts increase the dose to get the same high, digital addicts increase the amount of time they spend on devices.

You're already losing control.

In the early days you're able to convince yourself that you remain in control of when and how much you use devices and that there will be no problems.

But as time goes on and the feeling of pleasure or comfort grows more and more elusive, you begin to sense that you're slipping further and further into a bottomless pit. It's an unhappy, insecure feeling that creates further anxiety and stress.

Rather than recognizing that your digital dependence is causing this feeling, you assume that you're not getting the gratification because you're not going far enough! So rather than reversing the process, you actually increase your time on devices until you're completely consumed by them.

By now you've conditioned yourself to seek comfort from anxiety and stress by looking at your phone. So rather than dealing with the real cause of misery, you use the temporary fix of digital gratification to numb it.

You can see how this becomes a vicious circle of need, fix, withdrawal, need, fix, withdrawal… And because you keep increasing the dose, the highs become more short-lived, the lows more intense, and the net effect is an increasingly rapid descent, like the fly sliding into the belly of the pitcher plant.

This is how the trap works. It's how any addiction works.

DIGITAL ADDICTS SEEK COMFORT IN THE VERY THING THAT'S CAUSING THEM MISERY

> **NATURE'S WARNING LIGHT**
>
> Pain, whether physical or emotional, serves a useful purpose: It tells us that something is wrong. The solution is to identify the cause of the problem and fix it. But we're conditioned by modern medicine to take a different approach. We treat the symptoms, not the cause. Got a headache? Take a painkiller. Feeling anxious? Pop a Xanax pill.
>
> Imagine you're driving a car and the oil light comes on. What do you do? Remove the bulb from the warning indicator? Or pull over and top up the oil? Both actions will stop the oil light from flashing; only one will prevent the engine from seizing up.

AN INEVITABLE DECLINE

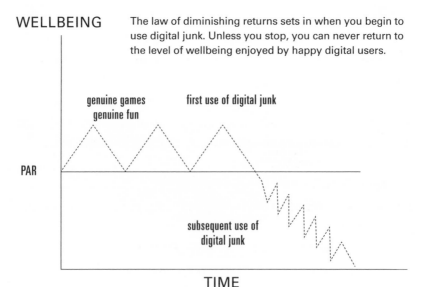

WELLBEING

The law of diminishing returns sets in when you begin to use digital junk. Unless you stop, you can never return to the level of wellbeing enjoyed by happy digital users.

genuine games
genuine fun

first use of digital junk

PAR

subsequent use of
digital junk

TIME

The nature of the trap is depicted by the diagram you see opposite. It is a graphic illustration of a digital addict's level of wellbeing in life. Remember, happiness is the sense of wellbeing that

What's the difference between REAL games and digital junk and online gaming? In the latter, defeat and victory are immediately disposable. There's no real element of jeopardy. It's trash – just press "Restart."

we get from genuine pleasures. such as playing real games or being with friends. False pleasures erode your happiness and wellbeing. Before you start, you enjoy genuine pleasures that take your level of happiness above par. When the feeling subsides you return to par. Of course, there are genuine lows in life, too, which take you below par, and when you get through them you return to par.

When you hijack genuine pleasures by using digital junk, you mistakenly think the feeling of genuine pleasure created by watching or playing sport, or taking on challenges, is replicated by the junk. But this is an illusion. The chemicals released by digital junk don't take you any higher, but the withdrawal from them takes you lower when they subside. Now your idea of par is lower than it was before as you feel the uneasy symptoms of withdrawal, like an itch you want to scratch. I call this withdrawal the Little Monster. It's so slight as to be almost imperceptible and it quickly passes. The Little Monster was created the first time you used digital junk and triggered the flood of chemicals. It feeds on those chemicals and, when you don't give it what it wants, it demands a fix. This too is barely perceptible, but the real problem is that it arouses another monster.

This second monster isn't physical but psychological. I call it the Big Monster and it's created by a combination of the brainwashing that has led you to believe that digital junk provides a pleasure or crutch. The Big Monster interprets the Little Monster's demands as "I need to check my phone/game," and so you end up trying to relieve a craving by doing the very thing that caused it in the first place.

So you scratch the itch—you check your phone or game—and you experience the illusion of pleasure. But it doesn't give you the genuine high that you knew as a non-digital addict; it gives the illusion of a high, which quickly wears off and the withdrawal takes you back down to a new low.

Perhaps you're thinking, "So what? Won't the boost I get from my phone or online game make me feel better, even if it's an illusion?" No, it won't! The reason you're reading this book is that digital junk is ruining your life, taking your wellbeing lower and lower. As time passes, you slide further and further down the scale and feel worse and worse.

As the body builds tolerance to protect itself from the chemical overloads, the quantities of chemicals you produce naturally have a diminishing effect. That's why addictions leave you feeling dissatisfied and unstimulated, more stressed, less able to concentrate, tired, lethargic, and miserable. The chemicals that normally keep these feelings in check are no longer having an effect.

Now you need a bigger "fix" to get the chemicals to work. For the drug addict, this means a bigger dose of the drug; for the digital addict, it means more and more digital junk.

The result is another false high and an ever-lower low. No matter how big a "fix" you give yourself, you can never return to par because of your body's increasing tolerance. As long as you keep relying on digital junk to give you the illusion of pleasure, your wellbeing will continue to fall lower and lower.

At the same time, your physical and mental wellbeing are also continuously being damaged in other ways. You stop taking care of your physical health and you become more and more conscious of the miserable trap that has imprisoned you. This combination of factors means that as you go through life as a digital addict, the lows get lower and the false high you come back to when you use digital junk goes down in proportion. By the time you're ready to accept that digital junk and online gaming are causing you nothing but harm, you're so far down in the trap that, like the fly in the pitcher plant, escape can seem impossible.

The good news is that when you quit with Easyway, escape is easy and, what's more, you can very quickly go back again to your original par and the genuine happiness of the non-addict. Remember, millions of people, who've found themselves in the same trap and been convinced that they will never be able to escape, have gotten free and so will you.

Like the fly in the pitcher plant, you only realize you're trapped when you're well and truly hooked. But there is one crucial difference between the pitcher plant and the digital trap:

IT'S NEVER TOO LATE TO ESCAPE FROM THE DIGITAL TRAP

Unlike the fly, you are not standing on a slippery slope; there is no physical force compelling you to use more. The trap is entirely in your mind. Ingeniously it makes you your own jailer, so the more you struggle, the tighter the bonds become, but fortunately for you that is also its fatal weakness. You have the power to set yourself free.

Further good news is that all those points on the scale are recoverable when you stop. Without the mental lows induced by digital addiction, you will be able to get back to par and start seeing the occasional low as a natural part of life, not something you need to relieve by looking at your phone.

Remember too that the fact that you've fallen into the trap has nothing to do with your character or personality. Millions of people who have found themselves in the same trap and been convinced that they will never be able to escape have gotten free and so will you, simply by following the instructions in this book.

SO WHAT'S HOLDING YOU BACK?

You should be able to see now that each time you turn to digital devices for pleasure or comfort, what you're really doing is trying to get back to feeling how you felt before you became hooked. You should also be able to see that there is only one way to achieve that: Stop using junk apps and games, and stop being pushed around by inappropriate social media use. But you also know that it is not that simple. There are two major myths keeping you in the trap:

1. The myth that junk apps, gaming, and social media excess gives you pleasure and/or comfort.

2. The myth that quitting them will be hard and miserable.

As long as you believe these two common myths, you will find it hard to quit. No matter how much you *want* to quit, part of your brain will be telling you you'll be happier if you don't. You need to change your mindset and that means unraveling these myths.

We are conned by Big Digital into believing that junk apps give pleasure and comfort. As you will discover later, there is no genuine pleasure or comfort to be gained from junk apps and games. There is plenty of misery and discomfort, as you have discovered.

But part of your brain will still be telling you that junk apps and games are a source of pleasure or comfort for you and you might be thinking, "Does it matter that it isn't true if it feels like it is?"

There are two very strong arguments that should answer that question once and for all.

1. Digital addiction is a threat to your physical and mental health, and that of your loved ones, and there's only so long you can go on burying your head in the sand before these grim realities take a very heavy toll.

2. If you were happy about your digital use, you wouldn't be reading this book.

Sooner or later, all addicts realize that there is no pleasure or comfort in their addiction. By this stage they are like the fly, descending rapidly into the belly of the pitcher plant. There is a point at which the fly senses that all is not well and thinks about flying out. It's nearly always too late. The fly struggles a bit, loses its footing, and falls into the digestive juices.

For you, though, there is no physical force preventing your escape. So when you sense that you are being consumed by digital overkill and want to get free, you are in a very strong position to do so. You know there is no pleasure, no comfort, no reason at all to keep using it. From here, it is a short step to establishing the mindset you need to walk free.

It's harder for people who still have the illusion of pleasure to change their mindset.

Once the illusion of pleasure is gone, there is only one thing holding you back: fear.

It's a strange kind of fear: The fear that escaping the misery of digital addiction will leave you in a more miserable place. But it's a very real fear and we will tackle it very soon.

First, though, we need to look more closely at the illusion of pleasure and the false beliefs that keep it alive in your mind.

Chapter 7

BUYING THE CON

IN THIS CHAPTER
•INSTINCT V. INTELLECT •HOW TO SPOT THE TRUTH
•I NEED MY DEVICES •FEAR OF MISSING OUT
•I CAN HANDLE IT •IT'S MY JOB—MY COMPETITIVE ADVANTAGE
•IT'S THE WAY I'M MADE

People who aren't hooked up constantly to their phone, emails, social media, and messenger apps find it hard to understand why you can't just put it down or disconnect. What they don't realise is that the trap is a clever combination of lifestyle, professional ambition, and a variety of illusions that make it impossible to see the way out. Easyway enables your escape by helping you to see through those illusions.

Addicts, including digital addicts, suffer a lot of negative emotions. One of these is a sense of foolishness. "Why do I keep doing this?"

When you know you're doing something that makes you miserable but can't resist the temptation to keep doing it, you can't help feeling foolish and helpless.

For a being as sophisticated and intelligent as a human, it does seem incredible that we can be so easily duped into self-destruction. And remember, it's not just you who has been taken in by these illusions;

millions of people have fallen for the same con and millions more are falling for it every day.

Much less sophisticated animals in the wild don't suffer from addiction, although scientists have succeeded in inducing addictive behaviors in laboratory animals. Addiction is a human condition.

Animals also use pleasure to identify the things that are beneficial to their survival. When they find food, they use their senses to test whether it is good to eat. They look at it, touch it, sniff it, and lick it, and only if it passes the test with these four senses will they eat it. Animals are designed to use their senses and instincts to survive. But so are humans. So why have we become a race of addicts when animals have not?

The crucial difference between humans and wild animals is that animals survive by instinct alone. We also use instinct to survive—it tells us when and what to eat, it alerts us to danger, it even helps us to find a suitable mate—but we don't rely on instinct alone. We also use intellect.

Intellect has enabled us to learn and pass on our learning, with the result that we have developed into a highly sophisticated species that is not only capable of dominating the rest of the animal kingdom but also of building fantastic structures and machines and appreciating art, music, romance, spirituality, and so on.

Intellect is a wonderful thing, but it can go to your head; we tend to respect intellectual behavior above instinctive behavior and this is what has led us astray. Instinct is Mother Nature's survival kit, but when it conflicts with our intellect we usually side with the clever camp.

WE THINK WE CAN OUTSMART THE SURVIVAL INSTINCT

A stark example is the sportsman who needs a painkilling injection in order to play. His instincts are giving the clear signal to rest and allow the injury to recover, but his intellect tells him he can numb the pain and play on. The result is irreversible damage to his body. His instinct was right, yet he chose to side with his intellect.

Look again at the "advances" the human race has made and you'll see that, rather than building on the advantage that Mother Nature has given us, we have devoted a remarkable amount of time to self-destruction. We have devised, and continue to devise, ever more sophisticated ways of killing each other in battle, but it's not just when we're trying to be destructive that our intellect leads us astray. It's also evident in the entertainments we pursue and the way we go about our work.

We have become a species of compulsive junk consumers by allowing our intellect to trick our instincts. And that junk includes the digital apps and games that you are hooked on.

A lot of the intellectual choices we make in life are beneficial. For example, intellect enables us to forecast and preempt problems in a way that other animals can't, and thus to protect ourselves better. But choosing to consume addictive junk is not beneficial. So why make that choice?

The reason is simple:

WE DON'T ALWAYS REALIZE WE HAVE AN OPTION

Of course, every time you pick up your phone you have a choice. Nobody is holding a gun to your head and forcing you to do it. But the brainwashing is so strong and the illusions it creates so convincing that it doesn't occur to you that you have a choice.

You are now aware of all the tricks that digital designers use to hook consumers into using their apps. These apps are the digital equivalent of refined sugar. They seem innocent enough and they feel like something you need and love, but they are actually bad for your physical and mental health and they give you no genuine benefit whatsoever.

Big Digital manipulates your instincts so that what begins as an intellectual decision—the first time you ever use a digital device—becomes instinctive. It feeds you with false information that overrides your natural instincts and fools you into believing the two myths:

1. The myth that junk apps give you pleasure and/or comfort.

2. The myth that quitting them is hard and will make you miserable.

HOW TO SPOT THE TRUTH

It's not hard to trick the human mind. The table diagram in Chapter 3 is a visual example of how easy it is to sow false information in your mind. But it also illustrates the fact that once you've seen through an illusion, you can't be fooled by it again.

This helps to answer a question that a lot of people ask of Easyway:

HOW DO I KNOW THAT EASYWAY IS NOT JUST BRAINWASHING ME IN A DIFFERENT WAY?

The way addictions work is by bombarding your brain with a false sense of pleasure and reward, which your mind comes to mistake for the real thing. Your instinct has been reprogramed into believing that you need to consume junk apps and be constantly online and connected in order to get pleasure or comfort. At the same time, your intellect has been fed the same misinformation. As long as you continue to believe intellectually that digital devices give you pleasure or comfort, you will not be able to reprogram your instinct to distinguish this false reward from the real thing.

Easyway does the opposite to brainwashing. It helps you to see through the illusions and unravel the myths that hold you back from walking free.

The first illusion is the one that most of us are taken in by.

I NEED MY DEVICES

Digital applications tap into our need to fill vacuums, our love of new things, our fascination with cause and effect, our desire for completion, our hatred of anxiety, our need to be liked, and our need for individual identity.

By massaging all these instinctive cravings, it creates a new fake need: the need for digital itself. This is an absolute dream for Big Digital. Thanks to their clever choice of bait, users are jumping on the hook. But just as refined sugar appears to satisfy our need for the sweet foods that are essential for our survival (fresh fruit and vegetables), this is a false satisfaction. It doesn't really fulfill our needs.

This is evident in the way we consume digital apps and comms. Let's compare it to food again. When you eat an apple, say, you eat at a measured pace, chew each mouthful, and when the apple is finished you feel satisfied. Now compare that to eating a chocolate bar. The pace of consumption is quicker, you don't savor the taste, you devour it as quickly as you can, and when it's finished you invariably want another one.

There can only be one reason for that: You want to get your fix as quickly as possible. It's not genuine pleasure that drives us to eat these foods; it's the *idea* of pleasure.

We don't eat them for the taste; we eat them for the sugar which manipulates and interferes with our natural urges. Sugar addiction is a huge issue in its own right and if you are concerned with your sugar intake (that includes starchy carbs), then check out my *Good Sugar Bad Sugar* book.

Now compare that to the way you consume digital apps and comms. Watch other people looking through their social media feeds or emails. Their movements are rapid, they scroll very quickly, stopping occasionally, then quickly moving on, all fast finger movements and twitchiness and occasional "fast thumbs" replies. This is not the way you behave when you're enjoying a genuine pleasure or engaging in something meaningful and constructive. It's all about getting as much as you can as quickly as possible, or getting through as much as you can as quickly as possible.

Next time you use your smartphone for anything—an app, a messenger service, or just emails—pay attention to the way you go about it. Slow down, focus on each post or message, and be aware of how it makes you feel.

You will become increasingly aware that you're consuming a lot of unsatisfying stodge.

Most junk apps and comms provide very little satisfaction at all. In fact, apps are programed that way because the designers know you will keep coming back in search of those occasional morsels of satisfaction.

But it's not the satisfaction that keeps you coming back; it's the desire for satisfaction and the belief that the app can provide it.

FEAR OF MISSING OUT

Even in the depths of despair, some digital addicts continue to believe their phone is the one thing that can make them happy. It never dawns on them that it is the cause of their misery.

We are brainwashed into believing that junk apps and comms will make us happy by both Big Digital and by one another. We share digital environments whereby you either get on board or you miss the boat. Want to feel included? Join the group.

Wanting to feel included is another way of saying fear of being left out. This is a very strong emotion among humans and it's one that Big Digital trades on mercilessly. One of the symptoms of digital addiction is a sense of isolation. If you believe that your phone is your connection to the world, and particularly your social world, you will continue to use it in your desire to avoid further isolation.

But when you hook into your phone, you isolate yourself further. You go into your own little world, blocking out the real world around you, even when that real world happens to contain the very friends you're so desperate to connect with! You've been conned into believing that you can enjoy a better social experience via your phone than you can in person.

What exactly is it that you might be missing out on? Gossip? Chit-chat? A picture of somebody's cat? Someone bragging about how marvelous their children are?

Every once in a while someone posts something that is genuinely entertaining or moving, but is this really what you're looking for with your constant scrolling? Or is it just that you need to satisfy the desire to know everything that has been said, because the thought that everyone else might know something that you don't makes you anxious? In other words, you are using social media to relieve the anxiety caused by using social media in the first place.

The fact is that while you're desperately trying to avoid missing out by living your life through your phone, you really are missing out on the reality that is going on around you.

You don't have to drop out of social media entirely; you just need to trim, in some cases quite severely, the number of people you're engaging with. The fact is, there will only be a handful of people you're connected to who you really need to stay connected to... and even then, if there was an emergency in their lives, they'd contact you directly.

There is nothing better than culling 30 percent of your Facebook "friends." In most cases, they're people who rarely share anything of interest, of use, or of relevance to you. Decluttering your friends' list is an important process and it's not unfriendly; it's just decluttering. What you are left with are close family and friends—people, who although they might post occasionally irritating or embarrassing stuff, don't impose too much social media junk on you and, on the occasions that they do so, are easily scrolled past without wasting any of your time. These are also people on whom you can rely to contact you directly in the event that there is some kind of time-sensitive communication requirement.

TAKE YOUR HEAD OUT OF THE SAND!

Remember, the reason you believe that you get some pleasure or comfort from consuming junk apps, junk social media, and junk comms is because, when you are not doing it you feel restless and uncomfortable. When you log on, the uncomfortable feeling is partially relieved. It feels like a boost, but it is only making you feel like a Happy Digital User feels all the time.

IT IS LIKE DELIBERATELY WEARING TIGHT SHOES JUST FOR THE PLEASURE OF TAKING THEM OFF!

I CAN HANDLE IT

Aside from the fact that this is a bizarre reason for choosing to do anything, the evidence is that you can't handle it.

You've been warned that excessive use of screens can be detrimental to your physical and mental health, and you probably feel the physical and mental ill effects on a regular basis—yet you've convinced yourself you'll be OK. You're not one of those sad people who gets completely hooked and can't detach. At least, that's what you thought. When you go to buy these devices and download the apps, no one talks about the threat of addiction and the misery that can cause.

We see children permanently hooked into their phones, scrolling and typing and ignoring one another, yet we still don't label this a problem. We see statistics showing an increase in depression and suicide among young people, coupled with an increase in cyber bullying and social isolation caused by phone use, serious eye damage, and yet many schools still encourage children to have phones with them all day.

The digital age is still in its infancy and we know from the example of smoking that it takes years before we wake up to the real dangers of our addictions. But you know from your own experience how digital addiction can tangle you up in its trap and keep you hooked, even though you wish you could stop. By the time you realize the real danger of junk apps and comms, it's too late and you're well and truly in the trap. The good news is that it's never too late to get out. Once you've seen the truth, it's easy to dispel the illusion that digital addiction is just a relatively harmless habit.

Taste freedom for a moment. I bet that you've forgotten the sheer pleasure and relaxation of turning off your phone, curling up with a loved one (or even on your own), and thoroughly immersing yourself in a movie, a sporting event, or a televised music event on TV. We've become so used to dividing our attention between something we love doing, such as watching a movie, and something that persistently butts into that pleasure in the form of alerts, buzzes, and our own personal twitching to check our phone screens that the simplest of pleasures has been lost to us. Try it. Tonight. Register what it feels like; within a few moments you'll be immersed in the movie, relaxed, laid back, and disengaged from the junk life.

IT'S MY JOB—MY COMPETITIVE ADVANTAGE

This is one of the most destructive elements of digital addiction. It's something so many of us have slipped into merely as a result of our natural desire for *maximum diligence*: Wanting *to do the best* for our department, for our business, or for our organization, or as a result of our *intense ambition*: to want *to be the best in our department*, or in our business, or in our organization. The former is viewed as admirable; the latter is viewed less charitably by many and in my experience successful business people show strong aspects of both.

There is nothing wrong with either diligence or ambition… just as long as they don't destroy you, your quality of life, or your relationships.

It's a question of perspective and timing. How did we start slipping into that dark hole of digital addiction via email? Well, it normally starts on our commute. It seems sensible to utilize the time on the train or bus home from work to finish off correspondence and handle anything arising. Our productivity increases proportionately. It's not like we get to leave our desks any earlier; it just helps take the pressure off the next day, especially when we do the same on our journey into the office in the morning.

But then we experience that terrible demon, "Mission Creep." We apply the same "working commute" logic to having a beer when we get home, to watching TV with our loved ones, to having dinner, and even to our bedtime routine. The same thing happens in the morning. As soon as we wake up, we start checking email and messages in between our various bathroom activities and breakfast (if we even allow ourselves that pleasure).

Where is our downtime? When do we get our reboot or recharge?

Whether our motivation is our natural desire to be the best worker that we can be or blind ambition, it's an insane existence that usually ends up in disaster; you break the most important machine on this planet, yourself!

The same goes for people who don't work in an office environment. A doctor or nurse finishes a shift at the hospital and, as soon as they fire up their smartphone, they're inundated with all kinds of correspondence and messages. They're already exhausted, but then feel obliged to work through it all, business and pleasure alike, before they even try to unwind and relax. The same occurs during break times; there is simply no "down time" any more.

Business has changed beyond all recognition in the past 20 years and it's high time that we, the people, pushed back a little. Twenty years ago, someone who posted a business letter might expect to receive a reply within a week or so. Today they expect a reply within minutes or hours. It's not just your boss or your co-workers or clients, though. Twenty years ago, someone interested in a career in your business, or a customer with a simple question would need to pick up the telephone or address and mail a letter in order to correspond with you. Now they hit a few letters on a keyboard, often writing incoherently, as a result of which they expect a reply almost immediately 24 hours a day, 365 days of the year.

You need to protect yourself, your quality of life, and your relationships from this utter madness. Don't misunderstand me: You don't have to abandon your diligence or ambition; it's just a question of looking after yourself, which is every bit as important an area of diligence and ambition as being contactable. You need to establish some limits, some boundaries, some controls.

IT'S THE WAY I'M MADE

Addicts of all kinds often put their addiction down to a flaw in their personality. It's either a weakness in their temperament (a lack of willpower) or a predisposition to addiction (an addictive personality). The implication in both cases is that the situation is beyond their control, an excuse that suits addicts very well because it removes the pressure to stop.

If this sounds absurd, that's because addiction is absurd. It drives us to make absurd decisions and to behave in absurd ways. Addicts know that their addiction is destructive and desperately wish they could quit, yet they lie and make excuses and try every deceit in the book to make sure they can carry on.

The problem is that the bulk of information we receive about addictions is that you *do* need willpower to quit and that there *is* such a thing as an addictive personality. The fact that this misinformation is put about by many reputable organizations that exist to help addicts only adds to its potency. Why would an organization that genuinely wants to help people put out information that serves to imprison them more deeply in the trap? The simple answer is because they too have been brainwashed and have never stopped to look at the situation another way.

The absurdity of addiction is the reason why you need to keep an open mind and follow the instructions—because the truth is often the complete opposite of what you assume to be true.

The belief that your addiction to digital devices is down to a flaw in your personality is a form of denial. Rather than accepting that you have an addiction and taking the necessary steps to overcome it, it enables you to say, "I have no choice but to carry on doing it."

But why would anyone want to say that? Why would anyone who is suffering the misery and slavery of digital addiction make an excuse that took away their option to walk free?

The answer can be encapsulated in one word, which lies at the root of all addictions:

FEAR

Chapter 8

FEAR

Addicts are caught up in a tug-of-war between the fear of what their addiction is doing to them and the fear they'll be worse off without the perceived benefits they believe they get from it. Let's examine your fears in more detail, so you can win the tug-of-war.

How often do you hear yourself making lame excuses? All addicts make lame excuses. That's how they are. And there never seems to be an end to it. Digital addicts have to make excuses because they know there is no logical reason to keep gorging themselves on junk apps, junk comms, junk games and junk social media. So what do you tell yourself?

"I'm just staying in the loop."; "I'll cut back tomorrow."
"I've started now, it'll be rude if I drop the conversation now."
"I'm just keeping on top of things."
"I might be tired, but this is keeping me going."
"I like online gaming—why shouldn't I do it every night? It's a social thing too, you know?"

"My online relationships are just as important as my real-world relationships."

"If my partner didn't moan so much about my online gaming, I might spend more time with her and less time doing it!"

You need excuses to stay hooked to your phone and gaming console because you're afraid of what might happen if you try to put them aside. You have been brainwashed into thinking that it gives you some pleasure, benefit, or comfort that you couldn't get otherwise and so the thought of life without it is scary. But at the same time, you're aware that it's making you miserable and stressed, exhausting you, making you feel like a pathetic, helpless slave, and deep down inside you're afraid of the damage this is doing to your relationship with your partner and kids (who you know deserve better), and to your mental and physical health in the long term.

You're caught in a tug-of-war between conflicting fears and it makes you think contradictory thoughts:

"I know it's making me miserable, but it's my one pleasure in life."

Think about that for a moment.

SOMETHING THAT MAKES YOU MISERABLE CANNOT BE A PLEASURE

PROJECTED FEARS

Human nature tends to make us cover up our fears. From an early age we're told "Be brave!" and we are taught to admire courage. Rightly so, courage is a virtue. But courage is not the absence of fear; it is the strength to act in spite of fear.

When you deny your fears you drive them deep, where they linger unchallenged. It's time to bring your fears to the surface and challenge them. It doesn't matter if you're not feeling brave—you can remove your fear of quitting with simple logic.

Fear is another instinct that is fundamental to our survival. It's nothing to be ashamed of. Fear is interesting because it can be both instinctive and intellectual. It is the instinct that drives us to fight or flight, alerting us to danger and making us wary in potentially dangerous situations. But the things that frighten us can be both real and imaginary. A fairground ride, a horror film: We know how to create fear without real danger; the suggestion of danger is enough to trigger the instinct of fear.

We can imagine danger, even where there is none. This is both an asset and a handicap. Our ability to learn about potential dangers also helps us to avoid them. The fears associated with losing your job, for example, are intellectual. You have learned about the possible consequences of finding yourself unemployed—e.g. having no money, being forced to sell your possessions, sacrificing the pleasures and comforts that you enjoy now, feeling worthless and unfulfilled—and so you do everything you can to safeguard your job and make yourself indispensible, even when there is no present threat of losing your job.

In this case your intellect does you a valuable service. But what if your projected fears are based on false information? You can be made fearful of dangers that don't exist too, like the fear of life without your digital fix.

There is absolutely nothing to fear from *not* using junk apps, junk social media, and junk comms and games. It was only very recently that the world existed without smartphones and tablets; nobody was miserable because they couldn't spend their day gazing at social media! Yet you've been brainwashed into believing that life without your regular fix will be empty and miserable.

As consumers, we are bombarded with so much false information that it's impossible for us to know what to believe. We end up spending a lot of our life worrying about things that will never happen—and being blasé about things that will.

Fear is the basis of all addiction. It is the force that makes the trap so ingenious, convincing the addict that there is some kind of pleasure or comfort in the trap. It is ingenious because it works back to front. It's when you are not looking at your phone that you suffer the empty, restless feeling. When you pick it up, you feel a small boost that partially relieves the restlessness and your brain is fooled into believing that your phone is providing a comfort.

This is why, as a digital addict, you cannot find true happiness while you remain in the trap. When you're on your phone, you wish you weren't. When you're not on it, you wish you were.

**ONCE YOU UNDERSTAND THE TRAP COMPLETELY,
YOU WILL HAVE NO MORE NEED OR DESIRE TO BE
HOOKED TO YOUR PHONE OR GAMES CONSOLE**

FEAR OF FAILURE

Being a digital addict is like being in a prison. Every aspect of your life is controlled by devices: Your daily routine, your hopes, your view of the world, your suffering. Of course, you're not physically imprisoned. There are no walls or bars. The prison is purely in your mind, but as long as you remain a slave to your phone, you will experience the same psychological symptoms as an inmate in a physical prison.

If you've tried and failed to conquer your problem, you will know that it leaves you feeling more firmly trapped than you did before you tried. It's like that moment in a movie when a prisoner is thrown into a cell and the first thing he does is run to the door and tug at the handle. This confirms his predicament: He really is locked in.

Trying and failing to unhook from your phone or gaming console has the same effect. It reinforces the belief that you are trapped in a prison from which there is no escape. This can be a crushing experience and it's quite natural to make your mind up that the best way to avoid the misery of failure is to avoid trying to get free in the first place.

The twisted logic of addiction concludes that as long as you never try to escape, you will always be able to preserve the belief that escape is possible. It is only when you try to escape that it becomes impossible. When it's written down like this you can see how absurd this thinking is, and yet it's not so clear when you're the one caught in the trap. The belief that escape is possible is very important to an addict. It represents hope. And who wants to risk shattering their own hope?

There are millions of intelligent people around the world who continue to keep themselves trapped in this way. They prefer to continue suffering the misery of addiction than risk the misery of failure. What they don't realise—and it's never pointed out to them until they discover Easyway—is that the person who tugs at the prison door and finds it firmly locked is using the wrong method of escape.

Tugging at the door is the equivalent of using the willpower method. The trap is like a snare, which tightens its grip the more you struggle. This is why it's often people with very strong willpower who find it hardest to quit an addiction—until they discover the Easyway "key."

The fear of trying and failing to quit is illogical. The thing you're fearing has already happened. You've become addicted. What could be worse? Every time you fall for the temptation to pick up your phone or ignore your partner to spend time online gaming into the small hours of the morning, you experience the sense of failure. You try to ignore it, but getting this far into this book is testimony to exactly that. As long as you remain an addict, you will continue to feel a failure. Perhaps you're afraid that trying and failing to quit will only increase your sense of failure. Rest assured, it won't—not if you follow the correct method.

When channeled properly, the fear of failure can be a positive force. It's the emotion that focuses the mind of the runner on the starting blocks, the actor waiting in the wings, the student going into an exam. Fear of failure is the little voice in your head that reminds you to prepare thoroughly, to remember everything you've rehearsed and trained for, and to leave nothing to chance. It can bring a remarkable clarity of thought and judgment.

But the addict's fear of failure is based on an illusion. The truth is you have nothing to lose by trying, even if you fail. The worst that can happen is that you remain in the trap. By not trying, you guarantee that outcome. In other words:

**IF YOU SUCCUMB TO THE FEAR OF FAILURE, YOU ARE
GUARANTEED TO SUFFER THE VERY THING YOU FEAR**

But there is another fear you have to unravel before you can make your escape.

FEAR OF SUCCESS

Many studies have been carried out into the psychological effects on prison inmates and ex-prisoners. One phenomenon that repeats with depressing regularity is the tendency for released prisoners to reoffend within a very short period after being let out. You might assume that these are habitual criminals who aren't very smart or have no moral judgment, but research has shown that in many cases ex-cons reoffend deliberately to get caught. They actually *want* to go back inside.

There's a tragic scene in *The Shawshank Redemption* when Brooks the prison librarian is released after serving a life sentence and finds that his life on the outside is meaningless, lonely, and degrading. He doesn't reoffend, he takes his own life.

Prison life is grim, but when it's the life you know, it can be more attractive than the alternative… or less frightening anyway. Life on the outside is alien and disconcerting. It's not what you know. You don't feel equipped to handle it. You yearn for the "security" of the prison.

This is very similar to the psychology of addiction. Addicts are afraid that life without their "crutch" will feel empty and pointless. They won't be able to enjoy its pleasures or cope with its stresses; they might even have to go through some terrible trauma to get free and then they'll be condemned to a life of sacrifice and deprivation.

We are all led to believe that life without our little indulgences is no fun. For digital addicts, this belief becomes the monster that keeps you in the trap. Though you're well aware of the misery that your addiction causes, you may now have come to regard it as part of your identity. Perhaps you've even convinced yourself that people like you or respect you because of it.

The constant use of digital devices is sold to us as the essential route to success, sociability, knowledge, and awareness. Without them, how can we stay in touch, both with our friends and what's going on in the world? People who don't buy into this are branded "technophobes" and you run the risk of being mocked if you're not up to speed with the latest news, gossip, or gaming news.

Perhaps you believe this is all true. Perhaps you believe that your phone, your online gaming, your complete immersion in social media keeps you in control, thriving, and aware of where the fun is. But ask yourself:

DO YOU FEEL IN CONTROL?

DO YOU FEEL LIKE YOU'RE THRIVING?

ARE YOU REALLY HAVING FUN?

Be absolutely clear about this: You have nothing to lose by escaping the digital trap. Life without the slavery of addiction is not something to fear; it is something to look forward to with excitement and elation. Choose to stay in the trap and you will feel a failure for the rest of your life. This is NOT the life you were born for. You have a choice.

WIN THE TUG-OF-WAR

The trap makes you your own jailer and this is both its fiendish ingenuity and its fatal flaw. The panic feeling that makes you afraid to even try to unhook from your phone is caused by using your phone. One of the greatest benefits you'll receive when you quit is never to suffer that panic again.

The tug-of-war is a conflict between two fears: the fear of what your digital addiction is doing to you and the fear of life without it. One of these fears is valid because it's based on fact; the other is invalid because it's based on illusions. But the tug-of-war is easy to win because both fears are caused by the same thing: using junk apps, junk games, and junk social media in a JUNK WAY!

THE WOW FACTOR

If you could be transported forward in time to the moment you finish reading this book, you would think, "Wow! Will I really feel this good?" Fear will have been replaced by elation, the feeling of failure by optimism, self-loathing by confidence, apathy by dynamism. As a result of these psychological turnarounds, your physical health will improve too. You will look healthier and you'll enjoy both a newfound energy, newfound vitality, and the ability to truly relax. Not only that, you will rediscover the ability to truly connect with your loved ones, connect with pure entertainment and relaxation activities, and most important of all, to connect with yourself, the real you.

Some people manage to go for weeks, months, even years without succumbing to the temptation to go back to their old ways on social media and junk apps but still find that they miss them. You may have been through this yourself. This method is different. Believe me, you will not miss the digital junk because you are not giving anything up. There is no sacrifice involved because junk apps, junk games, and junk use of social media do nothing for you whatsoever.

All you are doing is removing something from your life that makes you miserable, dull, bored, and boring, replacing it with something that makes you genuinely happy. It's no more difficult than throwing away a pair of horribly uncomfortable tight shoes and replacing them with a comfortable new pair.

YOU ARE NOT GIVING ANYTHING UP

You are trading lack of control over your use of digital for total control; no choice for absolute choice. Right now, part of you feels that digital junk is your friend, your constant companion, and comfort. Get it clear in your mind that this is an illusion. In reality, digital junk is your worst enemy and, far from supporting you, it's driving you deeper and deeper into misery. You instinctively know this, so open your mind and follow your instincts.

REMOVE ALL DOUBTS

Take a moment to think about all the good things you stand to gain by overcoming your digital addiction. Think of the enormous self-respect you'll have, the time and energy you'll save not having to make excuses, not having to manipulate situations so you can constantly game, or be on social media, or replying to junk comms. Think of the quality moments every day, when you can either lose yourself in your own thoughts, or call your parents, or meet up with your kids on the spur of the moment, or stay in the arms of your lover rather than feeling compelled to sit up and check your emails, or rush off to your gaming chair while they sleep.

That little boost you feel every time you reach for your phone is a mere hint of how you will feel all the time when you're free.

Can you think of anything more precious than time? Time to do things that really make you feel good. Feel incredible. Feel extraordinary. Or just feel genuinely relaxed and unplugged.

The difficulty for all addicts is being able to step outside their world and see their problem as other people see it.

If you saw a heroin addict suffering the misery of drug addiction, would you advise them to keep injecting heroin into their veins rather than try living without that high they get every time they get a fix? If you can see that the "high" is nothing more than blessed relief from the terrible craving that is caused by the previous shot of the drug as it leaves the body, you are well on the way to understanding your own addiction. And it should be clear to you that there is only one way to stop your craving.

STOP CONSUMING DIGITAL JUNK!

It's as simple as that. Once you can see that there is nothing to fear, that you are not giving up anything or depriving yourself in any way, stopping is easy.

So far, you've been given five instructions to put you in the right frame of mind so that this book can help you overcome your digital addiction:

1. FOLLOW ALL THE INSTRUCTIONS (CHAPTER 1)

2. DO NOT ALLOW YOUR DEVICES TO INTERRUPT YOU WHILE YOU ARE READING THIS BOOK (CHAPTER 1)

3. SEE YOUR SITUATION FOR WHAT IT REALLY IS (CHAPTER 3)

4. OPEN YOUR MIND (CHAPTER 3)

5. BEGIN WITH A FEELING OF ELATION (CHAPTER 4)

If you are struggling with any of these instructions, go back and re-read the relevant chapters. It's essential that you don't just follow the instructions but that you really understand them.

We have established that using digital junk does absolutely nothing positive for you whatsoever, that the beliefs that have imprisoned you in the trap are merely illusions, and that you have everything to gain and nothing to lose by quitting.

Can you see, therefore, that you have nothing to fear? If you can, then you're ready to move on.

If you're afraid that the process itself will be unpleasant, perhaps because you've tried to quit before and found it a torturing experience, remember that the willpower method doesn't work. It leaves you feeling deprived, which means you can never truly rid yourself of the desire to fall back into the trap.

Before you're ready to move on you need to be sure that you have removed the fear of success and the fear of failure. You need to be 100 percent certain about your desire to escape from digital addiction. If you have any doubts, please go back and reread this chapter, paying special attention to the arguments that prove that both fears have no logical basis.

SIXTH INSTRUCTION:
NEVER DOUBT YOUR DECISION TO QUIT

As you continue reading this book, you will be challenged to see things in a new way and this may cause you some doubts. It's absolutely fine to question what you read because that will help to reinforce the logic and truth behind it. But if you find yourself doubting your decision to quit, remind yourself why you chose to read this book in the first place and think about all the wonderful gains you stand to make.

If you're resolute about your desire to quit but still have doubts about your ability to do so, that's probably because you're not convinced that it's possible to quit without willpower. It's time we addressed this particular myth once and for all.

Chapter 9

WILLPOWER

There is a widespread assumption that any personal challenge cannot be overcome without applying all your willpower. When it comes to addiction, this assumption is counterproductive. While the determination to quit is important, the reliance on willpower actually drives you deeper into the trap.

We need willpower when forcing ourselves to go through pain or hardship. For example, if you run a marathon you will inevitably experience pain in your legs and lungs, but if you want to complete it in a good time you will push yourself through the pain. When one part of your brain is yelling "Stop!" but another part knows you need to push on, it's willpower that gets you through.

So how can you possibly conquer addiction without willpower? Simple: by removing the pain and hardship.

The assumption that overcoming addiction inevitably involves pain and hardship leads to the belief that willpower is essential for anyone who wants to get free from the trap. But quitting only involves pain and hardship if you believe that you are making a sacrifice. Easyway removes that belief and thus removes the need for willpower.

A common response to this claim is "If quitting is so easy, why do so many people find it incredibly hard?" The reason is simple: They are using the wrong method.

The simplest of tasks can become virtually impossible if you go about them the wrong way. For example, opening a door. You know how to open a door: You push on the handle and it swings open with minimal effort. But have you ever come across a door with no handle and pushed on the wrong side, where the hinges are? You meet with heavy resistance. The door might budge a tiny bit, but it won't swing open. It requires a huge amount of effort and determination to open it far enough for you to walk through. Of course, you don't keep pushing on the wrong side; you change your method, push on the correct side, and the door opens so easily you barely pay it any notice.

OVERCOMING ADDICTION IS JUST LIKE OPENING A DOOR. USE THE WRONG METHOD AND IT'S INCREDIBLY HARD; USE THE RIGHT METHOD AND IT'S EASY

Most digital addicts find it hard to unhook from their phone or games console because they try to use willpower to overcome the temptation. They have a constant conflict of will, a mental tug-of-war. On one side their rational brain knows they should stop because it's detrimental to their health and happiness; on the other side their addicted brain makes them panic at the thought of being deprived of their pleasure or comfort.

When you try to quit with the willpower method you set yourself up for a series of setbacks. You begin by focusing on all the reasons for stopping and hope your will is strong enough to hold out until the desire to go back to it eventually goes. This seems logical, but there is a problem: You still regard digital junk as a pleasure or comfort; therefore, when you stop surrounding yourself with it, you feel you're making a sacrifice. This makes you miserable. It also makes you feel you deserve a reward for your abstinence. Now, what do you do when you need cheering up or you feel you deserve a reward? You end up going back to your old behavior.

THE WILLPOWER METHOD HOOKS YOU MORE, NOT LESS

You only need willpower to stop if you have a conflict of will. You are going to resolve that conflict by removing one side of the tug-of-war, so that you have no desire to use digital junk. Using willpower for the rest of your life to try to resist the temptation is unlikely to prove successful and will not make you happy; removing the temptation altogether will.

Some people do manage to stop their addictive behavior through sheer force of will, but do they ever actually break free of their addiction? Are they ever truly happy? There are countless smokers who quit for years and then fall back into the trap. In fact, it's hard to find a smoker who hasn't tried to quit at least once in their life. There is no addict who genuinely enjoys their addiction and doesn't wish they could stop.

Despite having quit for so long that their addiction is a distant memory, these willpower quitters still believe that their little crutch is a pleasure or comfort. They haven't removed the desire and sense of sacrifice. All it takes is something to trigger their need for a pick-me-up or reward and they turn to their old fix again. Their will fails and they end up back in the trap, feeling more miserable than before.

HOW WEAK-WILLED ARE YOU?

Just as people generally assume that quitting requires willpower, those who fail to quit are assumed to be weak-willed. Indeed, addicts typically criticize themselves for being weak-willed and this can cause a strong feeling of self-loathing. The more you criticize yourself, the more worthless you feel; hence the less you try to look after yourself and the further you fall into the trap. If you think you've been unable to control your digital use up until now because you lack the strength of will, you haven't yet understood the nature of the trap you're in.

Strong-willed people actually find it harder to quit because they refuse to open their mind and accept that they're using the wrong method. They would rather wrestle with the problem than accept that they've been brainwashed and are not in control. The nature of the trap is such that the more you fight it, the more tightly ensnared you become.

FAILURE TO QUIT IS MORE LIKELY TO BE A SIGN
OF A STRONG WILL THAN A WEAK ONE

Addicts are stubborn. How would you react if someone tried to take your phone away, or told you you spent too much time looking at it, or if you came home and found your gaming gear locked away? Would it make you want to quit, or would it make you more determined to carry on, just to assert your will?

Ask yourself whether you're weak-willed in other ways. Perhaps you're a smoker or you drink too much, and you regard these addictions as further evidence of a weak will. There is indeed a connection between all addictions, but the connection is not that they are signs of a lack of willpower. On the contrary, they are more likely to be evidence of a strong will. Smokers, drinkers and other addicts cling stubbornly to their little crutch because they regard it as a symbol of their individuality and they don't like being told what to do.

When you consider that millions of people are imprisoned in the same trap in defense of their individuality, and kidding themselves they they are doing it through choice, you can see how absurd this attitude is. But addiction makes us think in absurd ways.

What all addictions share is that they are traps created by misleading information and untruths. And one of the most misleading untruths is that quitting requires willpower, because when you think something is going to be hard and unpleasant, you find excuses not even to try.

It takes a strong-willed person to persist in doing something that goes against their natural instincts. You know that the digital junk you consume is having an increasingly bad effect on your mood, making you stressed and unhappy, destroying your relationships, perhaps even your sex life, and leaving you feeling worthless and helpless; yet you keep finding excuses for doing it. That is not the behavior of a weak-willed person.

Addicts go to great lengths to cover up their addiction. Sneaking off to the restroom to check your messages, pretending you're expecting something important, feigning illness so you can stay at home gaming, putting your phone in your bag where you can see it every time it lights up... this level of subterfuge is stressful and requires a strong will.

If you saw someone trying to open a door by pushing on the hinges and you told them they would find it easier if they pushed on the handle, but they ignored you and insisted on pushing on the hinges, you would call them wilful, wouldn't you? You might call them crazy too, but certainly not weak-willed.

The prisoner who reoffends soon after being released from prison is not weak-willed; he is displaying a strong will to get back inside.

Think of all the people you know who seem to spend their lives on their phone or online gaming. It might be a small list because people go to great lengths to avoid appearing like they're hooked. But there are plenty of examples out there to illustrate that digital addiction is not exclusive to the weak-willed.

When you think about people who are successful in business, they tend to be heavy digital users. Some of them carry two phones! You don't reach lofty positions in business without being strong-willed. Yet when it comes to unhooking from their devices, they find it as challenging as you do.

When you try to quit by the willpower method, the struggle never ends. As long as you continue to believe that you're giving something up, you will always be running in pain. The stronger your will, the longer you will withstand the agony, but you never actually reach the finish line. In order to feel that sense of true escape, it is essential to remove the feeling of sacrifice.

CROSSING THE LINE THE EASY WAY

With Easyway, there is a finish line and you don't have to wait months or years to cross it. You cross the finish line as soon as you remove the fear and illusions and lose the desire to consume digital junk. That's when you are free of the addiction. You need to understand that you will not get to that line by forcing yourself through suffering.

Addicts do not respond well to a hard-line approach. Rather than helping you to quit, the willpower method actually encourages you to stay hooked because:

1. It reinforces the myth that quitting is hard and, therefore, adds to your fear.

2. It creates a feeling of deprivation, which you will seek to alleviate in your usual way—you will fall back into the trap.

Once you have failed using the willpower method, it's even harder to try again, because you will have reinforced the belief that you have a problem that is impossible to cure.

People who have tried the willpower method and failed will tell you they felt an enormous sense of relief when they first gave in. It's important to understand that this relief is nothing more than a temporary end to the self-inflicted pain. It's not a relief that makes you feel happy. No one celebrates falling back into the trap. You wake up hating yourself. "I'm a failure, a slave, a pathetic, weak-willed addict."

That first fix after you've tried to quit is not pleasurable at all, despite what others might tell you. They're confusing pleasure with the relief of ending discomfort.

OTHER QUITTERS

From time to time you come across people who are trying to unhook from their devices, apps, social media, or gaming by using willpower. You probably admire their determination and wish you could do the same. Think again! Remember what you have learned about the willpower method and see things as they really are.

Other digital addicts who try to quit by the willpower method can be bad for your own desire to quit. They either talk proudly about the sacrifices they're making, or they moan about them. Either way, they reinforce the misconception that quitting demands sacrifice.

It is important that you ignore the advice of anyone who claims to have quit by the willpower method.

NO SACRIFICE

You are following a proven method to free you from the misery of being hooked on digital addiction: A simple, logical method to unravel the brainwashing and remove all desire for digital junk. It's important that you understand you are not giving anything up. Once you can see this, you stop being a victim of some kind of tug-of-war. Without fear of what digital addiction is doing to you—no longer being addicted removes that fear—and without fear of what life will be like free from digital addiction—no longer being addicted means there is no longer fear of life without digital junk—there simply isn't any tug-of-war. It's gone. No battle. No toughing it out. Just freedom.

Without the tug-of-war there is no need for willpower. Take away the fear and there is nothing to tug against. It's easy.

People who quit by the willpower method are always waiting for the struggle to end, but with Easyway there is nothing to wait for. As soon as you dismantle the belief that digital junk is an essential source of pleasure or comfort, you remove the desire and cure your addiction.

This is a thrilling moment—the moment you realize you are no longer a slave. If you've followed all the instructions and understood everything you've read so far, you should already be feeling a sense of excitement and elation. You have taken a major step in solving your digital addiction and you can see where you are going.

You are regaining control and soon you will be free.

There is only one more potential hazard that could be holding you back from feeling like you're taking control. Addicts who try to quit and fail put it down to a lack of willpower. Addicts who try to quit and fail again and again put it down to something bigger: A flaw in their biological makeup, generally known as an "addictive personality."

Regardless of who you are, the only reason why anybody tries and fails to quit is because they are using the wrong method. Until you remove the desire, you will never be free from the temptation that drags you back into the trap. You have decided to quit with a method that works—a method that has worked for countless addicts who have tried and failed many times with the willpower method.

So let's take any thought of personality flaws out of the equation.

Chapter 10

THE ADDICTIVE PERSONALITY THEORY

IN THIS CHAPTER
•*ANOTHER MISERABLE EXCUSE* •*JUST ANOTHER MYTH*
•*A MULTIPLE ADDICT'S STORY* •*WHY ME?*
•*BORN WEAK?* •*ONE FINAL STATISTICAL POINT*

Some so-called experts believe that there is a "type" that is more prone to addiction than "normal" people, and that if you happen to be this type, there is nothing much you can do about it. The fact is, anyone can free themselves from addiction simply by following the instructions of Easyway.

Among the symptoms of digital addiction is a feeling of foolishness and weakness whenever you find yourself drawn to your phone or gaming console. The tug-of-war between wishing you could stop and fearing that you couldn't cope without it creates a constant state of friction, which is confusing, frustrating, and degrading.

To overcome these unpleasant feelings, you make excuses for logging on.

"There might be an important message."

"I'm only looking; I'm not posting anything."

"I'll just have a quick browse and then I'll put it down."

"It's only social media. Everybody uses it."

"I have to check in with the Fortnite gang or they'll be worrying about me."

These excuses are based on the false assumption that digital junk gives you pleasure or comfort, but digital addicts need these excuses to explain away their inability to resist the urge to stay online.

When you understand the nature of the trap you're in and how it controls you, it becomes clear that there is no basis to these excuses and you can't convince yourself to go along with them anymore. If you've understood the point that digital junk gives you nothing whatsoever —no pleasure, no comfort—then this is fine. You don't need excuses because there is no temptation.

But if you still feel the pull of temptation, despite knowing that your favorite excuses have been dismantled, you look around for another explanation. "I must have an addictive personality."

What does this mean?

In short, the addictive personality theory assumes that some people have a genetic predisposition to becoming addicted. No matter how much they try, there's something in the way they're made that turns them into addicts. It could be cigarettes, it could be heroin, it could be cake, it could be their phone; they're bound to become hooked on something and once they are hooked, they can't get free again. Their personality keeps them trapped.

Many addicts pounce on the addictive personality theory because it's a convenient excuse to stay in the trap. The security of the prison and the fear of success override their desire to quit and the theory gives them the excuse to stop trying. This appeals to them because:

• They believe their fix gives them pleasure or comfort.

- They perceive quitting to be hard.

- They're afraid they won't be able to cope without their little crutch.

Now ask yourself what you believe. If you go along with any of these beliefs, you need to go back and reread Chapters 7 and 8. It is essential that you understand and have no doubt whatsoever that:

- The fix gives you no pleasure or comfort whatsoever. It merely gives you partial relief from the craving caused by the previous fix.

- Quitting is easy when there is no conflict of wills.

- Life without digital junk will leave you feeling fantastic compared to how you feel now.

Relying on the addictive personality theory as a justification for staying addicted will only insure that you remain forever trapped, your suffering increasing, and your misery leading you closer and closer to despair.

JUST ANOTHER MYTH

The addictive personality theory came about because so-called experts studying addiction noticed certain patterns among addicts that seemed to suggest a common trait. These patterns included:

- Addicts who go on craving a fix years after quitting.

- Addicts who become hooked on multiple addictions

- Addicts who become much more seriously hooked than others

- Addicts who share personality traits

We have already established why some addicts continue to feel the craving long after they've quit. Addiction is a mental condition, not a physical one, and if you quit without removing the belief that your "drug" gives you pleasure or comfort, you will always feel deprived and will always have to fight temptation.

Multiple addictions are quite common—for example, digital addicts who are also smokers or gamblers, or heroin addicts who smoke and are heavily in debt. All these addictions are caused by the same thing but it's not the personality of the addict: It's the misguided belief that the thing they are addicted to gives them a genuine pleasure or comfort.

THE MISERY OF THE ADDICT IS NOT RELIEVED BY THE THING THEY ARE ADDICTED TO; IT'S CAUSED BY IT

IN HER OWN WORDS: KAREN

I started smoking when I was 14. That was my first addiction. By the time I was in my 20s I was drinking regularly too. At 22 I decided to quit smoking. I reckoned the booze was OK—all my friends were drinkers—but the smoking was becoming anti-social, not to mention costing me a fortune. To help me quit, I developed what you might call a candy habit. Instead of buying my pack of cigarettes every morning, I'd buy a couple of packs of candy. Every time I felt the craving for a smoke, I'd have a candy instead. It seemed like a sensible deal to me—a lot less money and a few more visits to the dentist instead of the threat of cancer.

But then I found myself eating more and more candy and not just that, I was craving all sorts of sweet things: cookies, cakes, chocolate bars... I put on a lot of weight. I had always looked at obese people and thought, "Why don't you just stop eating so much before you get to that state?" But by my 23rd birthday I was one of them, weighing in at 225 pounds, and I couldn't seem to do anything about it.

When I looked in the mirror I was disgusted by what I saw. I knew what I needed to do, but I just couldn't bring myself to do it. I was hooked on junk food. Then I heard that smoking makes you slimmer. It was the best news I'd heard for years, because I still craved cigarettes, despite my attempt to replace them with candy, and I was desperate to lose weight. So there I was, back on the cigarettes, still on the booze, and hooked on junk food. And I tell you what, the smoking didn't make me slimmer at all.

I had turned to all these things because I'd been led to believe that they would give me pleasure, relieve my stress, make me feel better, but they were all doing the opposite. I was miserable, I was stressed, and I hated myself. I was just a hopeless case, doomed to get hooked on anything addictive that came my way. It seemed obvious to me that there had to be something in the way I was made that made me so weak when it came to smoking, drinking, and emotional eating.

> Then someone gave me Allen Carr's *The Easy Way to Stop Smoking* book and I learned that addiction is not the fault of the addict; it's the fault of the addictive drug and the fault of society that fools us into consuming these addictive things and thinking they give us pleasure or comfort. This was a revelation. It made me see that I could quit smoking without needing candy or any other substitutes to help with the cravings because there would be no cravings. And having quit smoking, it made me see that I had the power to control my eating habits too. It just required a change of mindset, from believing that I couldn't live happily without junk food to realizing that it was junk food that was preventing me from living happily.
>
> By my 25th birthday, I had quit smoking, was drinking very infrequently, and had lost 75 pounds. I can honestly say I found it easy. Once I had let go of the belief that I was doomed to be an addict, I was able to unravel the remaining illusions and the bonds that had kept me hooked just seemed to fall away.

WHY ME?

So why do some people fall deeper into the trap than others? Why can one person have the occasional cookie and leave it at that, while another ends up eating the whole pack?

Doesn't that suggest that one has an addictive personality and the other doesn't?

It does point to a difference between them, yes, but there are numerous differences between people that can explain why one person's behavior differs from another's, and none of them has anything to do with their personality.

Our behavior is closely linked to the influences we are subjected to as we grow up: different parents, teachers, friends, things we read, watch and listen to, places we go, and people we meet. These are all part of the brainwashing and they will all have a bearing on how quickly we descend into the trap. People with time and money on their hands tend to fall into the trap faster because there are no obstacles holding them back.

If you believe that digital junk gives you pleasure or comfort, and each time you use them you feel the partial relief of the craving caused by the previous fix, then your belief will increase, your desire will increase, and your determination to use them more will increase.

IT IS THE BELIEF THAT DIGITAL JUNK GIVES YOU PLEASURE OR COMFORT THAT HASTENS YOUR DESCENT INTO THE TRAP

BORN WEAK?

Have you ever felt that you and other digital addicts seem to be a different breed from everyone else? Maybe that thought consoles you somewhat, especially if you're hooked on online gaming. It's like being part of an elite tribe—you come and go, drift in and out—but normally it's the same group, growing ever wider, that you interact with while you game.

You might share similar character traits: an unstable temperament, which swings between exuberance and irritability, a tendency towards excess, a high susceptibility to stress, evasiveness, anxiety, insecurity. Do you feel more comfortable in the company of other digital addicts?

Beware the temptation to believe that these character traits are evidence of a shared personality flaw, which has doomed you all to be digital addicts. The reality is that they are the *result* of digital addiction.

All addicts feel more comfortable in the company of similar addicts but not because they're more interesting, free-spirited, or fun. On the contrary, the attraction lies in the very fact that they won't challenge you or make you think twice about your addiction. Why? Because they're in the same boat.

In fact, perhaps there are times when you sense that you're online more than anyone else? Perhaps people have said something like, "Yay—I always know you'll be around for a game online," and perhaps that made you feel every so slightly worried. Even the guy who's online all the time thinks you're online all the time.

All addicts know that they're doing something mindless and self-destructive. If they're surrounded by other people doing the same thing, they don't feel quite so weak. It makes them happy to think that there's someone else in the same boat—stuck in the digital addiction trap.

The destructive feelings of weakness, helplessness, stupidity, and hopelessness are a terrible reality for addicts of all kinds. You will know these feelings for yourself every time you give in to the temptation to go online when you really wish you could stop. They are the chief cause of misery and they drive you back to your little crutch time and time again.

The good news is that once you're free from your addiction, you won't just be saved from all the unhealthy effects of digital overkill, you will also be liberated from the terrible impact it has on your character, your self-esteem, and your feeling of wellbeing.

ONE FINAL STATISTICAL POINT

The addictive personality theory is based on genetics. It assumes there is a gene that predisposes some people to addiction. On that basis, there should be a fairly consistent proportion of the world's population who are addicts.

But this is not the case. Take smoking for example, because it's the addiction that has undergone the most research over the longest period of time. In the 1940s, well over half the population and a massive 80 per cent of the U.K. adult male population were smokers; female smoking didn't peak until the 1960s and 1970s. Today the figure for adult smokers in the U.K. is just under 15 percent. A similar trend is evident throughout most of the Western world. So are we to conclude that the proportion of people with addictive personalities has fallen off a cliff in just over half a century?

While the number of smokers in the West has plummeted, the number in the East has soared. So what's happened there? Have addictive personalities migrated like some kind of virus to Asia, and vice versa?

Digital addiction and its related health risks are on the rise. But no one is putting this down to the addictive personality theory; they're blaming the sheer volume of addictive apps, games, and social media platforms and the increasing brainwashing and manipulation from Big Digital compelling us to consume them. In the face of this barrage, you need to keep a clear head and maintain a good understanding of the trap they're trying to lure you into. Remember, when you've seen through an illusion, you cannot be fooled by it again.

It's essential that you understand that you didn't become a digital addict because you have an addictive personality. If you think you have an addictive personality, it's simply because you got hooked on junk apps.

This is the trick that addiction plays on you. It makes you feel that you're dependent on your little crutch and that there's some weakness in your character or genetic makeup. It distorts your perceptions and thus maintains its grip on you.

The addictive personality theory is a serious threat to addicts because it reinforces the belief that escape is out of your hands and that you are condemned to a life of slavery and misery. This is a myth created by the illusions that digital junk gives you pleasure or comfort and that quitting is hard. See through the illusions, blow away the myth, and escape is easy.

Even if you did suffer from an addictive personality or an addictive gene, the fact is that Easyway makes it ridiculously easy to get free— regardless of it.

The slavery and frustration of digital addiction will soon be behind you as you continue to read through this book. Once you can see the situation in its true light, you'll wonder how you were ever conned into seeing it differently. Like millions of people around the world, you have been the victim of an ingenious trap. Recognize the trap for what it is, dismiss the idea of a flaw in your personality, and you will be ready to walk free.

Just keep an open mind and keep following all the instructions.

Chapter 11

GETTING HOOKED

IN THIS CHAPTER
•THE BRAINWASHING BEGINS •FORBIDDEN FRUIT
•WHY WE CARRY ON •TWO MONSTERS
•FACE YOUR FEAR •STICK TO YOUR GUNS

You should now be clear that your addiction to digital junk is not down to a flaw in your genetic makeup, nor a weakness in your character, but to a concoction of myths and illusions that give you a distorted perspective of reality and your situation. You are ready now to begin your escape, unraveling the illusions one by one, starting with the reason you got hooked in the first place.

If you have followed the instructions up to this point and kept your mind open, you should now understand why it is that some people become addicted to supposedly everyday things like digital junk and others don't.

From the day we are born, we all experience occasional feelings of emptiness, like a void. Most of us are fortunate, and it is initially filled by parental love, kindness, and affection. We experience the void to differing degrees, because we have different experiences as we grow up.

The void feels like an emptiness that needs to be filled. It makes you feel ungrounded and in need of comfort. The most common time of life to do this is in your teens, when hormones are playing havoc with your mind and body, and parents are becoming a pain rather than a comfort. Rather than being our heroes we begin to appreciate that they have normal human frailties and failings and that they no longer seem to be the perfect guides. You maybe experience pressure at school and slowly become aware of your place in society. During this time of huge transition, disillusionment and insecurity are rife and you tend to search for something else, other than a parent, to cling to. It is the time of life when most addictions begin.

But digital addiction, in general, is starting much younger. Like junk food, there is no legal age limit on consuming digital junk. Digital devices are given to children as toys, with the clear message: "These things will make you happy." When you're given a message like that from the people you trust most, you naturally take it on board.

Whether you're a digital native, someone who has grown up with a phone and a tablet in their hand, or whether you're someone who grew up with typewriters, VHS, and Kodak camera film, the addiction is the same'—the route into it might vary slightly.

These days parents use devices and games as "treats" and "rewards" to cheer their children up when they feel sad and incentivize them to be good. "You can play on your games console when you've tidied your room." The message is clear: "These things will comfort and reward you." A generation of children is growing up believing that digital junk is an essential source of pleasure and comfort. More than that, it's now arguably perceived as "a right" as pressured parents feed smartphones to ever-younger kids. I saw on the news last week, a crib-maker in London has designed a baby crib which accommodates a tablet. For a baby?

But it's not just children who are being brainwashed. As explained, there is a child in all of us that needs to be loved, that craves affirmation, that is fascinated by cause and effect, that is uneasy with emptiness and anxiety, that loves new things. Digital devices, junk apps, junk social media, and online gaming hook into all these emotional needs and give us a little boost or reward that keeps us coming back for more.

Of course, there are plenty of people who experience the same brainwashing but don't develop an addiction. Are they different from you? Or are they just the flies that have yet to land on the pitcher plant? They too have been brainwashed to believe that digital junk will give them pleasure or comfort; they just haven't yet felt the need. All it takes is a crisis to shake them emotionally, or perhaps another addiction like smoking to overcome, and they too will fall into the trap.

An aspect of digital addiction that sets it apart from all other addictions is that the brainwashing multiplies once you're hooked in. Because these devices and apps are a means of communication, they give Big Digital direct access to each user and enable a constant barrage of manipulative messages. It's direct, straight into your inbox, or onto your screen, or your search engine. At least the tobacco industry has to find ingenious ways of stalking its victims. Digital addicts are like lambs to the slaughter.

FORBIDDEN FRUIT

It makes perfect sense that when you feel sad or insecure you should try to cheer yourself up by whatever means you believe will work. The baffling question is: Why do we believe that we will get that pleasure or comfort from things we know to be harmful and generally unfulfilling?

When we're young our parents give us candy as treats, but they also tell us that they're not good for us. We grow up fully informed about the effects of sugary foods on our teeth and our weight. Yet we also grow up seeing our parents and other responsible role models indulging in these things, so any sense that they might really do us harm is removed.

At the same time, there is something about being warned away from something that makes it all the more enticing. Don't go into the haunted house! Don't lean over the railing! Don't put your head out of the window!... What's the first thing you do when your parents aren't looking?

THE WARNINGS AGAINST DIGITAL OVERUSE ACTUALLY INCREASE THE TEMPTATION TO INDULGE IN IT

You would think that all the knowledge we now have about smoking and the risks it poses to life would be enough to make all smokers quit or at least prevent the next generation of youngsters from falling prey to the addiction. Yet there are smokers who have had legs amputated because of their addiction yet still can't overcome their apparent desire to smoke.

TELLING AN ADDICT THAT WHAT THEY'RE DOING IS BAD FOR THEM IS NOT ENOUGH TO MAKE THEM QUIT

Addicts aren't stupid. They know very well that what they are doing is bad for them. The problem is that they can't quit *despite* knowing it. That's what makes them feel stupid and pathetic.

It comes back to intellect overriding instinct. You've been given mixed messages: You've been warned against digital overkill, but you've also been told that digital gives you pleasure and comfort. In fact, undoubtedly digital life DOES provide you with genuine pleasure and comfort. It's the junk aspects that drag you down and destroy your happiness. Then you see normal, respectable people hooked to their phones, so you suspect the warnings must be exaggerated.

"OK, so I might spend too much time looking at my phone, but it must be essential or else all these people wouldn't be doing it."

"OK, so Big Digital is turning into Big Brother, but it must be giving us something fantastic for so many people to take part."

Rather than take the warnings at face value, your intellect looks for a hidden message: "If people are doing it in spite of all the dangers I've been warned about, there must be something great about it." You want to know what that something is.

YOU DEVELOP A DESIRE TO GET HOOKED

The simple truth is that all those other people who make digital overkill seem normal have also been brainwashed and can't get off the hook.

It can seem to be a fine line between digital addiction or overkill and normal use; all you know is that you've crossed over it. The brilliant thing is that as a result of reading this book you're all set to step back over the line, redraw it, remember where it is, and understand how to insure you never stray over to the wrong side again. Ahead of you is a brilliant, bright future full of real fun, real relaxation, real pleasure, and real experiences. Digital life will cease to exhaust you and instead you will use it to save yourself time and effort while enhancing your everyday existence.

WHY WE CARRY ON

The next question is this: If there is no genuine pleasure or comfort in using digital junk, why do you continue to do it? Why, when you can feel that it's making you stressed and unhappy, do you not go back to the warnings and steer clear?

Why do you carry on no matter how bad it makes you feel? When you're in the back of a cab feeling nauseous because all you've been doing throughout the journey is look at the smartphone screen, and the motion of the taxi is making you feel worse by the minute, what do you do? Put your phone in your pocket? Of course not, you keep on scrolling and read on.

Do not underestimate the power of the brainwashing. You may not be getting the pleasure or comfort you expect from your digital use, but that doesn't mean you've seen through the illusions. Remember the table illusion in Chapter 3: Until you're told the tables are the same, your brain doesn't think to look at them in that way.

In the same way, the illusion that digital junk gives you pleasure or comfort remains intact, even when you can't feel it. The only way to derive satisfaction from your digital devices is to cut out the junk, but the brainwashing is so powerful that when you find you're not getting satisfaction, you assume that you need to use them more, not less.

THE REASON YOU CONTINUE TO USE JUNK APPS, JUNK SOCIAL MEDIA, AND JUNK GAMES IS THAT YOU ARE CHASING AN IMPOSSIBLE GOAL

That goal is satisfaction—the feeling of pleasure or comfort that you expect from digital devices. Digital addicts never feel properly satisfied. You only get satisfaction when you get what you've come for. Satisfaction is what a Happy Digital User feels all the time. The only way you can feel satisfied is to not use junk apps, not to engage in junk social media, and not to use online games.

What are the feelings that typically drive you to your phone, tablet or games console?

- Boredom—"It's something to do and it keeps my mind occupied."

- Sadness—"It helps me forget that I'm alone or unhappy."

- Stress—"It helps me to switch off and forget about my worries."

- Routine—"It's just what I do whenever there's a spare moment."

- Reward—"It gives me a little boost when I've been good."

- Opportunity—"I can't let the opportunity of being alone at home pass without maximizing my game time."

The last one is a real killer. Like an alcoholic who doesn't actually feel that he wants a drink but seeing as though there's an opportunity to have one—heck—why wouldn't he? These are not indicators of genuine pleasure. You must have had times in your life when you had a hobby that gave you genuine pleasure.

Say, you played tennis. People who love tennis would play every day if they could. They wouldn't wait until they felt bored, sad, or stressed. They would break their routine for a game of tennis. And they don't feel they have to earn the right to play.

They actively pursue the sport because it gives them genuine pleasure.

Next time you pick up your phone for pleasure or comfort, look more closely. Are you really doing it for pleasure or comfort, or are you doing it because something in your mind is compelling you to and making you feel uncomfortable if you don't? Examine the illusion. Can you see through it?

TWO MONSTERS

The word addict has always conjured up an image of the squalid drug addict. Even smokers struggle to regard themselves as addicts until they understand the nature of the trap they are in. Addict is an ugly word that no one wants as a label, so we use terms like smoker, gambler, and drinker instead. The latest addition to that list of names is "gamer."

But it is becoming increasingly evident that all these behaviors are addictive. Whether they involve a substance, like nicotine, alcohol, or sugar, or a behavior, like digital overuse, they take control of your brain in a very similar way to cocaine, heroin, and other hard drugs. There is a physical effect and then there is a psychological effect.

The physical effect is like a Little Monster, so small as to be almost imperceptible. It is a feeling of restlessness and emptiness, like a niggling itch that needs to be scratched. It's this feeling that makes you reach for your phone and become twitchy when you can't. Most of the time your immediate reaction, regardless of where you are, or who you're with, or what you're doing is to pick up your phone and unlock it. The same goes for gaming. How often do you voluntarily avoid going online when you have the opportunity? Does it ever happen?

Now, I'm not asking about the times when your partner is fed up to the back teeth with you ignoring them, or you ignoring the kids in favor of your online games… and that guilt, shame, or embarrassment motivating you to, albeit sulkily, do something else instead. When was the last time you sat at home, with free access to your games console, with no one else around, when you chose to watch a movie, or have a long bath, or read the newspaper, or call your parents rather than go online?

The Little Monster was created the first time you used digital junk. When it doesn't get its fix it begins to complain. The feeling is very slight, but it is dangerous because it arouses another monster—the Big Monster in your brain.

The Big Monster is not physical but psychological. It is created by all the brainwashing and it interprets the Little Monster's complaints as "I need my phone" or "I need to game." So you end up trying to satisfy a craving by doing the very thing that caused the craving in the first place.

Every time you look at your phone or go online to game it quietens the Little Monster, creating the illusion that the device has made you feel better. In fact, all it has done is taken you from feeling restless and twitchy to feeling slightly less so.

Before you created the Little Monster you may have felt restless and twitchy from time to time, but the feeling wasn't permanent. You didn't need a stimulant to just feel OK. Now you need it again and again just to stop feeling miserable.

With the cycle of addiction, you never get back to how you felt before you got hooked—not until you're free. Look again at the diagram in Chapter 6. Every time you give yourself a "fix," you develop a tolerance against it. So every time you consume junk apps you need to consume more to get the same boost, and every time you stop you sink lower. The longer you go on trying to satisfy the Little Monster with junk apps, the lower your mood sinks and the more dependent you feel.

This is why your use of digital never leaves you feeling fulfilled.

MINDLESS SCROLLING

Have you ever picked up your phone to carry out one simple task, such as sending a text message, and found yourself ten minutes later scrolling through some other app or newsfeed? Don't you find it happens all the time? Big Digital has designed it, so that as soon as you look at your phone you are bombarded with temptations to flit around looking at other things. You don't want to miss out, so you slavishly click the links, telling yourself, "I'll just have a quick look while I'm on —it'll save time later."

Most of us have trained ourselves to avoid the pointless clickbait, but every now and again we get taken in by it.

It's the equivalent of opening a box of chocolates with the intention of only eating one and finding yourself ten minutes later having devoured it all. You feel bloated and sick, yet unsatisfied, disgusted with yourself, and horribly out of control.

Why do we do it? There is no pleasure in it. The more you consume, the less you notice what you're consuming. It just becomes a race to look at as much junk as you can as fast as you can. Only when you're interrupted can you stop. Clearly you're not scrolling through this stuff for the quality of the content, you're devouring it to satisfy the greedy Little Monster. There is nothing to savor. The lack of satisfaction drives you to the next link and the next and the next and the further you go without achieving satisfaction, the more frenzied and ridiculous the scrolling becomes. Like a dog chasing its tail, you go faster and faster, chasing something you can never attain.

The low you feel as you experience a feeling of withdrawal (the Little Monster) is compounded by the low brought on by the psychological craving (the Big Monster) and as you slide further into the trap, the realization of your predicament brings you down even more. This triple low becomes your new idea of normal.

Unlike the Little Monster, the Big Monster really can make you miserable. When it is awakened, it fills your head with the illusory feelings of deprivation, reminding you of all the false ideas you've been fed about junk apps, junk social media, and junk games being a source of pleasure and comfort, and compelling you to get another "fix."

The only "pleasure" you get is the mild relief of the feelings of withdrawal. In other words, the Little Monster is kept quiet for a while, but you know that when it awakes again, its cries will be louder than ever. And the Big Monster's influence will be stronger than before.

"What are you missing out on among your Facebook friends?"

"What's happening in the WhatsApp group?"

"What are my online gaming pals up to?"

"Are they missing me? Are they wondering where I am? Do I count anymore?"

Instead of responding to the Big Monster by saying "Nothing" or "No" or "Who cares?!" to each of these questions, we imagine that we're missing out on some tremendous happenings and events. In truth, the digital junk world is where nothing meaningful, nothing genuinely pleasurable, nothing real ever occurs. Yet we still cave into it like a junkie being peddled low-grade, cheap, killer junk.

This is the cycle of addiction that keeps you in the trap, even when you know you're getting no pleasure or comfort from each fix.

> **FIX FIXATION**
>
> It's revealing that junkies use the word "fix." A fix is a solution to a problem; you fix something that is broken. When something breaks and you fix it, it is never quite as good as it was originally. A fix is not an improvement. It is not a pleasure or reward. It is nothing more than partial relief, like taking off tight shoes.

FACE YOUR FEAR

All digital addicts wish they could feel like Happy Digital Users. But you are afraid to "give up" your little pleasure or comfort. This conflict leaves you feeling helpless and stupid. Why can't you just take control of the situation and sort yourself out?

Unfortunately, the usual response to this confusion is for addicts to bury their head in the sand and pretend that they don't have a problem. They lie to themselves about the state they're in, get angry and resentful with anyone they perceive as audacious enough to draw their attention toward the problem, and pretend that all they're doing is treating themselves to a bit of harmless indulgence. As long as you keep your head in the sand, you will not be able to see through the illusions and you will remain in the trap, suffering increasing misery every time you give in to the temptation.

The beautiful truth is that you can start feeling like a Happy Digital User any time you want to. All you have to do is stop using digital junk. It's as easy as that, provided you understand the trap you're in and follow the right method to get out of it.

You've already taken a big step. You have overcome denial and accepted that you have a problem. That's why you're reading this book. Now all you have to do is kill the Big Monster. Once the Big Monster is dead, you will find it easy to cut off the supply to the Little Monster and it will die very quickly and painlessly.

You kill the Big Monster by removing the illusions that create the desire to reach for your phone. Think about how those illusions were created and by whom The friends and colleagues who are all hooked to their phones just like you are have all been brainwashed themselves. Big Digital, which continues to peddle that message of pleasure and comfort, has a vested interest in you remaining hooked.

Don't give it the satisfaction. You have a right to happiness and it is standing in your way.

STICK TO YOUR GUNS

Your brainwashed belief in the illusion of pleasure is what put you in the trap. Now that you understand the nature of the trap, you can see that there is no genuine pleasure in using junk apps, just an illusion that drags you into a downward spiral of misery. You understand that junk apps do not fill the void; they make it bigger. Your brain is unraveling the illusions.

YOU HAVE ALREADY BEGUN TO KILL THE BIG MONSTER

Now we need to make sure that you are not diverted from your course. There are still several things that could sabotage your escape plan. There may be a part of you that still believes you get some pleasure or comfort from digital junk.

Deep down inside, though, you know the key areas that feed the problem, don't you?

It's too many Facebook friends... depending on how far you're in the pit of despair you need to get rid of many of them. How many is normal? Maybe 70 to 90. That includes friends, family, colleagues who you are friendly with, ex-colleagues, and people you meet as a result of any hobbies you might have managed to maintain (watching or taking part in sports, or quiz nights, or whatever).

Any more than 90—in all seriousness—and some need to go. Don't worry about them being offended. If you're worried that they might be, by all means send them a message saying you're trying to trim down your social media and definitely don't want to lose touch with them. But if you're honest with yourself they rarely, if ever, interact with you on social media and won't even notice that you're no longer Facebook friends.

Too many WhatsApp or Snapchat groups? How many do you really need to be in? Do you really need to be discussing the daily gossip related to the latest reality show on TV, or your baseball team's latest signing, or what everyone is doing at the weekend? There's nothing wrong with a few groups of very, very close friends, and very close family. You know what the junk is, so get rid of it.

Online gaming? Really? Are you a child? There are plenty of people who do too much of something and become fanatics—baseball fans are a good example. But whereas their fanaticism might take up their Saturday afternoon or Wednesday evening, you have gotten yourself into a position where most of what you do is online. Gaming and chit-chatting with people who really don't count, do they? Well, you seem to think so, which is why you'd rather spend every spare minute with them in a weird online, phoney fantasy world rather than live in the moment with the people you really love.

You know it's weird, and odd, and not right—that's why you're reading this book. Unlike social media, or messenger groups, it's not a case of trimming things down—it's a case of getting rid of it all.

By all means, if you value some of the friendships you've made online, then stay in touch with those people. But talk to them one-to-one. Chat to them about life. Real life. Get out of the game, and stay out of the game.

If you're worried that without being "in the game" you won't have anything in common with them, well then, there's your answer. They mean as little to you as you mean to them.

Perhaps you are afraid that life without them will leave you feeling isolated and deprived. There are many influences out there that will mislead you with false ideas like these and some of them are well-meaning. Please take note of the next instruction:

SEVENTH INSTRUCTION: IGNORE ALL ADVICE AND INFLUENCES THAT CONFLICT WITH EASYWAY

The simple truth is you are not "giving up" anything. You are freeing yourself from a trap that has been making you feel like a miserable, pathetic slave. Rejoice! The Big Monster is already dying and soon you will be ready to kill the Little Monster too.

Chapter 12

CHECK YOUR DIGITAL HOURS

IN THIS CHAPTER

•WHY YOU LOOK AT YOUR PHONE •FEELING SATISFIED

•DIGITAL OVERDOSING

•WHAT'S GOOD FOR YOU

The brainwashing has caused you to develop a relationship with your phone and other devices, apps, and gaming platforms that is not healthy. But you can unravel the brainwashing very quickly and restore a healthy control of your digital use just by challenging the Big Monster and questioning the beliefs that have led you into the trap.

Do you ever stop to ask yourself why you use your phone? Not why you use it the way you do, but why you use it at all? The obvious answer is that you would be unable to communicate if you didn't. But is that what you're thinking whenever you pick it up?

Next time you feel the urge to look at your phone, ask yourself why you're doing it. On most occasions the answer would be that you have no specific purpose; you're just doing it as a subconscious reaction to an urge. That urge is the Little Monster, and the subconscious reaction is the Big Monster.

THE FACT IS, YOU'RE NOT REALLY PAYING ATTENTION TO WHY YOU'RE ON YOUR PHONE

Routine, boredom, restlessness, stress, and anxiety are the common triggers. They have nothing to do with staying in touch. You look at your phone to fill the void and to divert your attention from other problems. To fill gaps in your life that might otherwise be vacant, still, and peaceful.

You have a difficult problem at work so you reach for your phone. It looks like you're doing something constructive; you even convince yourself that that's the case. In truth, you're just avoiding the problem. Looking at junk apps might take your mind off your anxiety for a moment, but sooner or later you have to return to the problem and it hasn't gone away. In fact, it's becoming more of a problem the longer you leave it unchallenged. Your phone doesn't take away your problems, but as long as you regard it as a comfort, you will reach for it every time you feel anxious or every time you might otherwise be silent, processing your feelings and thoughts without interruption or influence.

Our phones have become ridiculously useful. Need to know what time the next train is? The phone! Want to know when the team you support's next home game is? The phone! Want to know who sang the song you're listening to in a bar? The phone! Need to know what you're doing tomorrow? The phone!

We rarely use our phone specifically to de-stress, but we often kid ourselves that's one of the benefits.

The other reason is the cycle of addiction. The feeling of withdrawal from the previous fix adds to your anxiety. When you use digital junk you partially relieve the feelings of withdrawal, creating the impression that the device or app has eased your anxiety.

FEELING SATISFIED

Your smartphone is an incredible tool. It contains an extraordinary number of useful applications that can really make life easier and more enjoyable. When you pick up the phone for this purpose—say to order a taxi, find your way to your destination, translate a message in a foreign language for work, check the weather, reserve a table at a restaurant, or make an online bank transfer, whatever—and you use it only for this purpose, you feel satisfied when you finish. As you can see, there is a huge number of nonjunk apps and non-digital junk resources that you should always feel happy to use.

It's like eating healthy food as opposed to eating junk food. Junk food contains none of the useful vitamins or minerals your body needs, so it doesn't satisfy hunger. Junk apps are the same: They are not designed to provide a useful service; they are designed to keep you hooked in for as long as possible, so you never feel satisfied.

SO. YOU. NEVER. FEEL. SATISFIED

Satisfaction is the feeling that tells us when to stop. Without it, we find it very hard to stop. We crave that cause and effect, we crave that sense of completion, we crave the reward, the affirmation. As long as we are kept craving these things, there is always the feeling of "just one more," hoping that "one more" will bring the satisfaction we crave, the satisfaction that is designed never to come.

Next time you look at your phone, and its junk apps, pay attention to how it keeps you hanging on. Ask yourself if it's giving you a sense of satisfaction. There will be small rewards, just enough to keep you hooked, but do you come away with a feeling of completion, like when you finish a job?

Or are you already feeling the urge to pick your phone up again as you're putting it down?

You reach for your phone to fill the void, but the way you use it leaves you with a bigger void. The only way to close the void again is to stop using junk apps. Use your phone as a tool that does the job you want it to do and you will find it satisfying. You will also find it easy to put down.

DIGITAL OVERDOSING

If you're in the routine of looking at your phone at certain fixed times, such as coffee breaks, in bed at night, and first thing in the morning, then you're not using it as a useful tool and you are bound to use it to excess.

In fact, you're probably laughing at the suggestion that you might only look at your phone at those times. The fact is, it's like a constant, nervous tic, every few minutes unless you're in a situation where you can't possibly check it. OK, you might go without looking if you're at a movie or the theater or having a nice meal out, but even in those circumstances you probably have to resist the urge to check your phone repeatedly. You wouldn't use a screwdriver according to a fixed routine, would you? Or a hammer? Your phone has many more uses than a screwdriver or hammer, but the point still stands: If you use it regardless of whether you have a genuine need, you will not come away satisfied.

When you overdose on digital junk, it feels like bingeing on junk food. You're left feeling physically and mentally sick. You feel sluggish and gross, you lose self-respect, and regard yourself as weak and pathetic. You blame yourself for your lack of self-control and you make yourself feel utterly miserable. You might react by denying yourself for a while. After a period of abstinence, you feel you've paid your penalty and you deserve a little reward. You reach for those junk apps again.

And the cycle of misery repeats. The fatal flaw in all this is the misconception that you lack the control because you're weak-willed. As you now know, it's nothing to do with willpower.

YOU LACK CONTROL BECAUSE YOU ARE NOT IN CONTROL. ADDICTION CONTROLS YOU

When you understand this crucial point, it changes the game completely. You can see that your task is not to summon up more willpower; it is to cure your addiction. One requires a massive effort; the other is easy. All you have to do is follow the instructions.

Pay attention to your feelings when using digital devices and you will notice when you're getting satisfaction and when you're not. Your natural instincts will start to regain control over the intellectual desire to keep chasing an unachievable goal.

This will help you to avoid overdosing.

WHAT'S GOOD FOR YOU

You may be thinking this all sounds rather austere. Phones are supposed to be fun, aren't they? Well, that's up to you. They can certainly give you pleasure by helping you do things quickly and efficiently. They can give you comfort too, by putting you in touch with the people you love. You can use them for entertainment, such as watching a film, reading a novel, or the newspaper, or listening to music, podcasts... the list goes on and on.

But, do you really want to play games on your phone? Really? Do you think that's a constructive or beneficial use of your time?

OK, aside from the blue light, a game or two of solitaire before bed might be restful, but if you're playing every spare moment you have, it's causing you a problem.

In order to make that decision so it benefits you, you need to be completely in touch with your feelings as you sort through your apps. And you need to remember that Big Digital has a vested interest in hooking you in any way it can.

As digital users we don't stop to ask ourselves why we're reaching for our devices and we don't ask ourselves whether we're really enjoying the experience. As a digital addict, you can feel that you're not and now you should have a good understanding as to why that is.

Your senses haven't died; they've just been sidelined by the brainwashing. You'll be amazed how quickly they leap back into action when you give them some attention. And when you listen to your senses, you'll begin to unravel the brainwashing very quickly too.

Up until now, you haven't really been exercising your own choice over your digital use because you've been brainwashed into choosing what Big Digital wants you to choose. By bringing your senses back into play, you'll be able to make a genuinely informed choice about what your favorite apps are—a choice informed by your own personal detection system.

You have come a long way already and you are well on your way to becoming a Happy Digital User. Congratulations! It is natural to have some questions and lingering doubts at this stage, however, and it is essential that you remove all doubt about your decision to take control.

In the next chapter we will address the questions that typically arise at this stage, clearing the way for you to complete your escape.

In the meantime, why not check your notification settings? Do you really need to hear from any of your apps? Get rid of as many apps and notifications as possible—not the junky ones you keep using—but start clearing out the dead wood and start getting rid of those notifications. Guess what, if someone shoots the president, you'll find out about it sooner or later and you really don't need to know about the latest pizza deal or addition to your podcast library, do you?

Chapter 13

DIGITAL DILEMMAS

Soon you will be ready to make the simple changes that will free you from your digital addiction and make you a Happy Digital User. First, though, we need to make sure there are no lingering doubts that might undermine your decision to quit.

Everything you've read so far has been aimed at changing your mindset: from believing that digital junk is an essential source of pleasure or comfort to understanding that it actually creates and adds to a void.

You should be clear now that your problem is not due to any weakness in your character, be it a lack of willpower or an addictive personality, nor to anything special about the apps, social media, and games you use, other than the fact that they are designed to addict you and keep you hooked. You should also be well into the process of reassessing the apps and social media you need and those you don't, recognizing the difference between useful apps and junk apps. Paying attention to your feelings each time you look at your phone will help with this.

If you're reading this book because your digital addiction revolves around online gaming, then it's understandable to have feelings of uncertainty. After all, I've already said you need to ditch the games entirely rather than sort the "useful" from the"junk." Is that unfair? No. Whether you're addicted to stuff on your phone like Candy Crush, Angry Birds, Need for Speed, Pokemon Go, or games adapted for phones like GTA and Fortnite, or whether you're into what you consider to be more serious online gaming via console, laptop, or PC, these games have blurred the lines between fun and entertainment and addiction to such an extent that addictive qualities are ingrained and intertwined in their DNA.

All online games are deliberately designed to be addictive to maximize the playing time of anyone engaging with them. There is not a game corporation boardroom in the world where the question, "How do we make our games less addictive?", is being seriously discussed.

Oh, for sure the subject "How do we make our games APPEAR to be less addictive?" is talked about at length, but with an insincere objective like that it's akin to tobacco giants Phillip Morris International declaring that they want to help to cure the world of smoking. Believe it or not, they're actually uttering those words, but would you believe them? Or would you understand immediately that they're up to something?

> ### A CHEROKEE LEGEND
> #### *THE BOY AND THE RATTLESNAKE*
> A little boy was walking down a path and he came across a rattlesnake. The rattlesnake was getting old. He asked, "Please little boy, can you take me to the top of the mountain? I hope to see the sunset one last time before I die." The little boy answered:

"No, Mr Rattlesnake. If I pick you up, you'll bite me and I'll die." The rattlesnake said, "No, I promise I won't bite you. Just please take me up the mountain." The little boy thought about it and finally picked up that rattlesnake and carried it up to the top of the mountain.

They sat there and watched the sunset together. It was so beautiful. After sunset, the rattlesnake turned to the little boy and said: "Can I go home now? I am tired, and I am old." The little boy picked up the rattlesnake and, holding it tight against his chest, he came all the way down the mountain and took it home, where he gave it some food and a place to sleep. The next day the rattlesnake turned to the boy and said: "Please, little boy, will you take me back to my home now? It is time for me to leave this world, and I would like to be home now."

The little boy carefully picked up the snake, held it close to his chest, and carried him back to the woods to die. But before he laid the rattlesnake down, the rattlesnake turned and bit him in the chest. The little boy cried out and threw the snake to the ground. "Mr Snake, why did you do that? Now I will surely die!" The rattlesnake looked up at him and grinned, "You knew what I was when you picked me up."

NO ESCAPE

Online gaming is the crack cocaine of digital addiction and it has been proven to take control of people's lives in an intrusive, destructive, harmful way that never lets you rest or escape.

THAT'S EXACTLY WHAT IT WAS DESIGNED TO DO

If cutting down was the answer, addicts might have achieved that already—no doubt they've tried many times. It might work for a short period, but eventually your game time creeps back to where it was before, and possibly even beyond that.

The more you attempt to limit something you're addicted to, the more precious it seems to be. When you eventually crack, the more you gorge yourself on it and return to unfettered use. The great news is that you're going to find it easy to stop being used by these games; escape is easy when you follow the instructions.

If you have any lingering doubts about any of these aspects, please go back and read the relevant chapters again. Remember to keep an open mind and listen to your instincts, not the Big Monster in your brain.

It's essential that you achieve the right mindset to enable you to escape from the trap without a difficult struggle or any feeling of deprivation. .

When you started reading this book, you were probably eager to know how long it would take to work. How long would you have to wait before you could call yourself cured?

When you quit with the willpower method, you can never be sure that you've succeeded. It is a lifelong struggle to resist temptation. With Easyway, you know with absolute certainty when you've succeeded because you can feel that the temptation has been removed.

REPLACE UNCERTAINTY WITH KNOWLEDGE

Different people reach this stage of the book in different frames of mind. Some are confident that they understand everything and feel certain they are ready to unhook from junk digital straightaway. If that's you, wonderful, but let's not jump the gun. You may be convinced that you are in control and have no desire to ever use your phone in a pointless, twitchy way, and maybe you're champing at the bit to escape from the treadmill of constant online gaming and social media check-ins, but beware! The trap has ways of catching you out just when you least expect it.

It's important that you complete the book, so that you can walk free with all the protection you need to make sure you never fall into the trap again.

If, on the other hand, you have reached this stage in the book still uncertain that you have what it takes to overcome your addiction, don't worry. We still have some ground to cover and all will become clear. Whatever your current frame of mind, take the time and care to read right to the end of the book.

We need to make sure that you completely let go of any lingering belief in the willpower method. Remember, reliance on willpower actually makes it harder to quit and more likely that you will fall back into the trap.

Yet the brainwashing is so incessant that it can take a while to shake off the belief that overcoming digital addiction requires a very strong will.

There is nothing stupid or unusual in the belief that willpower is essential for your escape. The yearning you feel when you try to cut yourself off from junk digital may go against all logic, but the feeling is still very real and so is the irritability and misery you feel when you try to use willpower to deny yourself.

You've been subjected to the brainwashing all your life and you remain surrounded by it—if you've ever tried to quit any addiction by using the willpower method, you will have reinforced the belief that it's hard.

ANY FAILED ATTEMPT TO CONTROL AN ADDICTION THROUGH WILLPOWER REINFORCES THE BELIEF THAT IT IS DIFFICULT TO QUIT

When you're convinced that willpower is the solution, you blame yourself for your failures. This adds to your misery and low self-esteem and drives you deeper into the trap.

Free yourself from any belief in the willpower method and all of a sudden this vicious circle goes into reverse. Stop pushing against the wrong side of the door, open your mind to there being an easy way out, and that easy way will open up to you, as if by magic. Of course, it's not magic, it's simple logic—a logic people are blinded to by all the brainwashing. But when you open your mind and let the truth in, it can feel miraculous—a moment of revelation.

The truth is very simple:

- **The desire comes from the Big Monster—the illusion that digital junk gives you pleasure or comfort.**

- **The anxiety you feel when you can't look at your phone is merely the Big Monster responding to the cries of the Little Monster to be fed.**

- **The Little Monster was created by using digital junk in the first place.**

- **Therefore, digital junk does not relieve the anxiety; it actually causes it.**

Kill the Big Monster and you remove any feeling of deprivation when you put aside your devices. The willpower method is all about fighting through the anxiety for long enough until you no longer feel it. In the first few days after quitting, when your willpower is at its strongest, you will have the upper hand in the battle. But over time, as you believe you're winning, your willpower will slacken off.

They say a soccer team is most vulnerable just after scoring a goal. It's the same with the willpower method. It's when you think you're winning that you're at your most vulnerable. You can't maintain the effort it took to quit and your motivation to do so lessens. Then one day you notice the Little Monster crying to be fed and the Big Monster wakens with a roar. It's very hard to dig in again and you feel victory slipping from your grasp.

Now your mind is torn in two—one half determined to stay off the junk, the other urging you to indulge. Is it surprising that people get so confused, irritable, and downright miserable on the willpower method? It would be a miracle if they didn't!

If you're worried that you might fall back into the trap at some point after you've quit because that's what you've done before, remember that it is the Big Monster that pulls you back in. You have all the ammunition you need to kill the Big Monster. Pay attention to your senses and needs, stay aware of all the tricks being used to hook you in, clear your mind of the illusions, see the true picture, and you will walk free with absolute certainty that you have no desire to seek pleasure or comfort in junk apps again.

WILL I BE ABLE TO ENJOY GOOD TIMES?

One of the big selling points of social media is that it enables you to capture and share the fun moments in life. Parties, birthdays, vacations, and so on. A photo, a comment, a few shares and likes… it all adds to the fun, doesn't it?

Or does it? Did you not have fun before social media came along? Every second you spend taking a selfie, writing a caption, sending and then checking to see who has responded and what they've said is actually a second spent outside the moment. Try putting your phone away and living in the moment. Don't misunderstand me. Keep taking those wonderful photos, but make sure you enjoy the moment rather than frantically posting them to your social media. There's no hurry. They'll be just as gratefully received by close friends and family the next day or at the weekend. Creating memories doesn't mean filing your life on a social media profile; it means using your senses to soak up what's going on around you: sights, sounds, smells, tastes, and feelings. And there's nothing wrong with supplying your Facebook page or other platform with wonderful photos from the events in your life… it's like a never-ending photo album.

If you look at any big sporting event or musical performance or even movie premieres, all the spectators seem to be living the experience through their smartphones. It never occurs to those engaged in this madness that it might be more rewarding and exciting to live in the moment and look at the whole picture rather than focus on holding a smartphone up to capture the moment.

And what are these people left with? Video and photos of someone famous, a famous (or fairly run-of-the-mill) sporting moment, or a blurry musical performance. The kind of thing you can see anywhere online at the click of a button. There's nothing wrong with grabbing a selfie with a hero if you're lucky enough to meet one—it's the 21st-century equivalent of an autograph—but for your sake, and as a sign of respect to the person you want a selfie with, try to connect with them properly before expecting them to pose for the camera.

LIVING IN THE MOMENT

I've been extremely lucky to meet some of my all-time heroes in person. In all cases they were fortuitous, entirely separate "bumping into" moments rather than being the result of being in places where I might have been rubbing shoulders with the rich and famous. Muhammad Ali, David Bowie, and former Chelsea FC soccer players Peter Osgood and Ron "Chopper" Harris are the main ones that spring to mind.

My memory of meeting them all, chatting to them for a few moments, and shaking their hands is as clear to me now as it was the first time I excitedly told someone about it. Do I wish I had a photographic record of each meeting? Well, I guess that would be nice, but I wouldn't swap the privilege of having had a few moments chatting to them for a photograph. Could I have had both the chit-chat and the photo? I don't think so. It's all about having a few short moments to either connect—even on the most basic of levels—or grab a photo.

As for Muhammad Ali—arguably the greatest sportsman of all time—I look back at a few minutes of friendly chat with him as one of my fondest memories. I have no doubt that the coming together, conversation, and therefore the memory would have been less significant and special to me had I been attempting to squeeze him into a selfie (not that such things existed at the time).

John Dicey – co-author, Allen Carr's *Smart Phone Dumb Phone*

This book is all about helping you put the pleasure back into your life. The problem has been that you've been seeking pleasure from your phone, junk games, and engagement with social media in a junk way. You should now be clear that the way to achieve satisfaction is to avoid digital junk.

But when you've been a digital addict, bingeing on junk apps, games, and social media in the belief that they are an essential source of pleasure or comfort, you might be concerned that living the rest of your life without your usual "fixes" will be a bit miserable or boring. The truth is that your addiction actually reduces your ability to derive genuine pleasure or excitement from anything. If you believe you can't be happy without your digital junk, you won't. Another drawback with the willpower method is that you never properly remove this belief.

There are two important points to bear in mind:

1. Once you're back in control of your digital use and you're not turning to devices for emotional reasons, exposure or access to digital junk won't be a problem. You won't be interested in it and junk apps can only hook you if you use them.

2. When you no longer have the desire for digital junk, you don't regard it as a treat. You don't miss it because it makes no contribution to your enjoyment. In fact, it never did.

If you can think of occasions in your life that you consider to have been enjoyable because you had your phone to capture it on, analyse them carefully to understand why the phone appeared to enhance the situation and see that in reality it was more likely to do the opposite. Instead of perpetuating the illusion that such occasions won't be enjoyable again without always having your phone in your hand, take the opposite view: Remind yourself that you will now be able to enjoy those occasions more because you'll be free from the slavery of digital addiction. You're not going to stop taking photos of happy occasions —that would be crazy—but you're going to be enjoying the moments rather than acting like an unpaid event photographer.

Most of the time you're not even aware of how you feel while you're using your phone. The only time you're really aware of how it makes you feel is when you want it but can't have it, or when you're hooked into it but wish you weren't. In both cases it makes you miserable. The conclusion is obvious:

TAKE AWAY THE DIGITAL JUNK AND YOU REMOVE THE MISERY

WILL I BE ABLE TO ENDURE BAD TIMES?

This question is possibly even more important. Addiction is not just a pursuit of pleasure or comfort; it is an attempt to escape from unhappiness and discomfort. You have an argument, there's pressure at work, you're in financial difficulties, you suffer a bereavement... or you just feel bored, lonely, nervous... By burying yourself in your phone you kid yourself that you can take yourself out of the situation and put your problems out of your mind.

But sooner or later you have to return to the real world and the problems are still there. In fact, they've usually gotten worse. If you believe that your phone provides comfort in these situations, you will leave yourself vulnerable after you've quit. The next time you feel the need for comfort, the Big Monster will pull you back into the trap.

Think about it: How is playing a game on your phone or console going to help you with any of the problems in your life? Have you ever found yourself in the middle of a domestic row thinking, "It doesn't matter that we're shouting horrible, hurtful things at each other because I can just go and play that game and it will be all right"?

Or did the fact that you spend too much time on your phone or online gaming make the argument worse?

All you have to do is accept that you will have ups and downs in your life after you've quit and understand that if you start yearning for your devices in such situations, you'll be moping for an illusion, something which makes things much worse—not better.

DIGITAL ADDICTION REDUCES YOUR ABILITY TO COPE WITH STRESSFUL SITUATIONS BY ADDING TO THE STRESS

Even when you're completely clear about this, it's possible to feel confused when hard times do occur and you find yourself feeling low again. That familiar feeling can make an addict think they're falling back into the trap.

You can avoid this potential pitfall by anticipating the difficult times that will inevitably occur in life and prepare yourself mentally. Be ready to remind yourself that any stress you feel is not because you've unhooked from your phone.

Tell yourself, "OK, this is tough, but at least I haven't got the added problem of being a slave to my devices. I am stronger now." You will find that the stressful situations in your life will actually feel less severe once you are free from digital addiction.

BEING FREE ENHANCES ALL SITUATIONS IN LIFE—GOOD TIMES SEEM BETTER AND BAD TIMES ARE NOT NEARLY AS BAD

THE MOMENT OF REVELATION

Soon you will be ready to apply the practical steps to freeing yourself from your addiction to digital junk. There will be a final moment when you put your phone aside and vow never again to rely on it for pleasure or comfort and confirm that it is a functional tool. It can provide genuine pleasure, but it is not the cure-all, or panacea, for every situation, good, bad and indifferent. The same goes for social media. Later in the book, if you have online gaming issues, I'll ask you to do the same—to confirm to yourself that you've put them behind you forever. After this, you will be able to say you are no longer addicted to digital junk; you have no desire for it; you will be free from digital addiction.

The purpose of this moment is to draw a line in the sand: the point at which you walked free from the trap.

When you've been trapped in a prison for a long time and then the door suddenly swings open, you may sit and savor the moment for a while before walking out. That final vow is the act of walking out of the prison, discovering that your perception of the world around you really has changed and beginning your new life of freedom.

When you achieve something big, like passing an exam, landing a job, or winning an award, you experience a wonderful high as the realization of your achievement sinks in. A lot of people feel the same thing when they quit with Easyway—a moment of revelation. The confusion caused by addiction is suddenly replaced by absolute clarity and understanding—a thrilling realization that their desire for junk has gone.

This clarity is essential. It is not enough to *try* or *hope* that you will never get hooked again; you must be certain. Easyway is designed to give you that certainty.

Unlike the willpower method, Easyway doesn't leave you waiting for some sign of confirmation that you are free. You won't spend the rest of your life suspecting that there could be bad news lurking just around the corner.

YOU KNOW YOU'RE FREE THE MOMENT YOU EXPERIENCE THE MOMENT OF REVELATION

Chapter 14

TIME FOR ACTION

IN THIS CHAPTER
• GET REAL • PAY ATTENTION
• PERSONALIZE IT FOR PROGRESS
• IGNORE EMAIL • DISABLE PUSH NOTIFICATIONS
• OBSERVE HAPPY HOUR EVERY DAY

Now that you have reversed the brainwashing and you know that digital junk is not a source of pleasure or comfort, you're ready to start taking the practical steps to unhook from your devices. This chapter contains seven more instructions. All you have to do is follow them.

Take a pen and paper and write down all the activities that used to give you the most pleasure before you became hooked on digital devices. Really think about it. Take a day or two over it if you like and write down the things you enjoyed as they come to mind.

Focus on the times when you felt most relaxed, at your happiest and most contented, when digital addiction was the last thing on your mind. This will help you to become aware of what you really enjoy and value in life. Your list might look something like this:

• Seeing friends
• Going for a walk alone, or with a loved one

- Going for a bike ride
- Having a meal out with a loved one
- Reading a book
- Going to the movies or visiting a gallery, exhibition, or museum
- Playing or watching a sport
- Gardening or sitting in the garden
- Listening to music... listening to it and not doing anything else
- Playing an instrument
- Making love on a weekend morning
- Being cuddled up on the sofa watching a movie
- Reading a newspaper—a real one—made of paper!
- Kicking back and enjoying watching the world go by
- Daydreaming

The following practical steps will put you in control of your digital use, so that you can enjoy these genuine pleasures again.

EIGHTH INSTRUCTION:
GET REAL ABOUT YOUR DIGITAL USE

Do you know how much time you spend looking at your phone and/or gaming online? Have a guess. Two hours a day? More than that? Less? Nearly all users underestimate their daily phone use by a significant amount. Digital addicts underestimate by as much as half!

So if you think you spend 90 minutes on your phone each day, the chances are you spend three hours.

Online gamers who are forced by family life to resist game time during "normal hours" tend to go online at around 9 p.m. and finish at perhaps 2 a.m. as many nights as they can get away with—some even longer.

It's important to recognize the extent of your digital use. All addicts play down the severity of their problem because the alternative is the daunting prospect of having to do something about it. But you *have* chosen to do something about it so there is no need to kid yourself anymore.

You know how much time you spend on your phone or online gaming and you know you want to escape from it. You're nearly there.

NINTH INSTRUCTION:
PAUSE AND PAY ATTENTION

It's time to say goodbye to mindless digital use. As instructed in Chapter 5, you should be compiling a sorted list of useful apps and junk apps, useful/essential WhatsApp/Snapchat/Other groups and junk ones. Each time you look at your phone and other devices, pay attention to your feelings. Ask yourself why you're looking, what you're looking for, and how long you intend to spend looking.

Don't be afraid to challenge yourself in this situation. Ask yourself questions like:

"Do I really need to look at my phone now?"

"Who am I doing it for?"

"What am I expecting to gain?"

"What will give me satisfaction?"

"Why am I part of this group? Is it really necessary or desirable?"

Simply asking these questions will introduce a pause in what would otherwise be the headlong, mindless "rush to the front door to see if the mail has arrived". This could be enough to divert your mind toward a different decision, which in turn could save you the disappointment of finding all those "brown envelopes." At the very least, it will put a check on the impulsive, mindless use that is symptomatic of digital addiction. It will always make you more aware of the amount of time you're spending on the device, and the distractions that are leading you off from your original purpose.

Being more aware of your phone use will also help you to recognize the logical reasons for controlling it, which then makes it much easier to do. This is the basis of Easyway. Once you see the logic, you can remove the desire and it becomes easy to quit.

Use this new awareness to make a note of how each app you use or group you're part of makes you feel: good or bad, stimulated or numb, satisfied or frustrated, distracted or focused? Use this to complete your list of useful apps/groups and junk apps/groups. This list will determine which apps you choose to keep and which you choose to get rid of.

"BUT I GET GREAT CHAT ON MY HOCKEY WHATSAPP GROUP!"

Well, that might be the case, although if you analyze the historical chat you'll probably realize how utterly banal it is most of the time. Consider the cost of engaging in this group? Of course it's not financial. But if you're engaging with that group, who are you neglecting as a result? Loved ones? Work? Other friends? There's nothing wrong with being in a few groups, but be picky. Less is best.

TENTH INSTRUCTION:
PERSONALIZE IT FOR PROGRESS

We all like to personalize our phones: how you arrange your apps, the picture on your home screen, the sounds it makes for each type of notification. Personalization is one of Big Digital's tricks for hooking you, but you can use it to turn the tide and help you with your escape.

Start with your home screen. Instead of setting it up with some enticing image of a tropical beach that makes you want to dive in (notice how the default images that your phone came programed with are all beautiful and aspirational), choose something that reminds you to stop and think. A question mark, say, or a statement like "Pay attention!" Your home screen is the gateway to the digital world of wonders where Big Digital wants you to spend the rest of your life. Use it as a reminder of the dangers that lie within. Over the coming weeks, this will act as an instant reminder to think twice about checking your phone. Once you feel completely free, there's absolutely nothing wrong with changing your home screen back to your favorite photo again—whether that's a beach shot or a family pic.

The next thing to do is arrange your apps to suit you and your needs. Referring to your list, remove all junk apps, keep any borderline ones as far from your home screen as possible and have only useful apps on the home screen: for example, phone, messages, map, calendar, settings. If there are other useful apps that you use regularly, such as contacts, clock, and so on, try to place these on your home screen too. The less you have to swipe to find the things you need, the less likely you are to be distracted by things you don't need.

ELEVENTH INSTRUCTION:
IGNORE EMAIL

Email is a useful thing but it is also one of the most intrusive of all apps. If you can do so, turn off the alert and move the app from the home screen. The problem is that it delivers new messages every few minutes, so whatever you happen to be doing, you feel obliged to stop, read, and reply. This is hugely distracting, wastes a great deal of time, and can be immensely stressful.

Instead, treat your email like a really good "snail mail" service that makes three deliveries a day. Schedule a time for emailing in the morning, at lunchtime, and near the end of the day. This will enable you to see anything that needs your attention during the day, catch anything that may come in mid-morning, and then tie up any loose ends before you finish work for the day.

If you have a separate email account for personal emails, apply the same technique. Only log on at certain times. Once a day should be enough for personal emails. Remember, if people need an answer from you urgently, they can phone you.

For the rest of the time, ignore it. That means when you're at home too. Some countries are already imposing restrictions on the use of company email after hours. They have recognized not only the need to curb the intrusiveness of email but also to curb the demands of employers who use the omnipresence of digital media to keep their staff "at work" all hours of the day, every day of the year. This is not only an abuse of the employer–employee relationship, it results in bad work. It's often not even the employer's fault as diligent employees strive for ever-greater productivity. We voluntarily take our desk home with us on the train or bus and set it up in our living room, bedroom, and even bathroom.

A scheduled approach to email will enable you to do all your emailing much more efficiently, because you'll be doing it in blocks, rather than in dribs and drabs. Your mind will be able to focus on one activity at a time, which will make you faster at processing your messages.

If you worry that switching off your email will make you miss something important, put it to the test. Try it for a couple of days and see how many emails you receive are obsolete because you didn't read them immediately. It is highly unlikely that there will be any. If there are, ask yourself just how important the thing was that you missed out on. And how likely is it never to come round again?

Despite our reliance on email, when something is absolutely time-critical we still pick up the phone. Think about it. If someone you know needed urgent medical attention, would you send the hospital an email? Of course you wouldn't. Make it a rule that if somebody needs to get hold of you urgently, they can call you. Don't regard email as a medium for instant responses. It was no more designed for that than snail mail was.

If the last few paragraphs have infuriated you because you have a job or business that requires constant vigilance regarding emails and messages, or you receive so many emails on a daily basis that your work consists almost entirely of email handling, you have my deepest sympathies. Advice such as "only check emails three times a day" simply doesn't apply to you. But, you can do something to ease the strain. You must.

Twenty years ago, even relatively junior managers would have access to a typing pool. The more senior you became, the closer you got to having your own PA. The route normally included sharing a PA with a few other managers. Of course these days we'd probably find it impossible (and actually harder) not to handle our own correspondence from receipt to reply. If things had stayed the same, it wouldn't have caused so many problems. But things never stay the same.

The volume of written correspondence has increased many times over, not just ten or twenty times more. No one needs to handwrite a note, address an envelope, attach a stamp, and mail anything anymore. All it takes is someone to find an email address and off they go. Any query, no matter how mundane or how pointless, can land in your inbox. Often it comes to you via a colleague's inbox. It's becoming almost impossible for businesses—large or small—to cope with the volume of written correspondence, much of which is quite pointless and unnecessary. Suddenly you're not only taking all the big decisions in your department or business or organization, but you end up dealing with endless reams of minutiae, trivia, and frivolous enquiries or feedback.

GET HELP!

At some point in all this madness you need to make a little room for yourself. Don't forget that 20 years ago you might well have had your own PA helping to take the strain. Don't let your personal bandwidth stretch so much that it breaks you. This is how burnout happens.

Address the volume of emails now. It's not something you can fix overnight, but it is something you can plan for and handle over the coming months. Be wary of letting more tasks and projects barge into your personal bandwidth as you free up space in it. This is a real discipline and one performed by every successful person on the planet —well, those that live to tell the tale, of course.

TWELFTH INSTRUCTION:
DISABLE PUSH NOTIFICATIONS

Any app that you have on a mobile device can send you messages, regardless of whether you're using the app at the time. These are called push notifications—an apt name because they push you into taking actions you wouldn't otherwise take.

The impression is that you are being constantly updated with news and information that keeps you in the know. The real purpose, though, is to enable app publishers to keep you on the hook by tapping into your love of new things and your fear of missing out. Every time your phone makes a noise, you can't resist looking to see what the notification says. You're running to the front door, only to find another brown envelope waiting on the mat. Even if you don't open it, you've let it break your rhythm.

Although you have to opt in to receive push notifications, we often do so automatically as soon as we download the app. Often this is because in that moment we do want the app to notify us, but this makes them a very powerful marketing tool and a major obstacle to your escape from the trap. The beauty is you can turn them off. Permanently. It's as simple as that.

In the settings on your mobile device, you will find a section for controlling apps and notifications. You can disable push notifications app by app, or you can disable them all at once. It takes about ten seconds and it will relieve you from the constant jostling, which is one of the most intrusive, distracting aspects of digital devices.

If you struggle to find out how to do it, google it—tech can be so useful sometimes.

THIRTEENTH INSTRUCTION:
OBSERVE HAPPY HOUR EVERY DAY

Happy Hour is the hour before you go to sleep, spent doing things other than looking at screens. It makes you happy because it enables you to get a good night's sleep.

The blue light emitted by screens fools your brain into thinking it's still daytime, which suppresses the release of the sleep hormone melatonin. If you spend time looking at a screen right up to the moment you go to sleep, your brain will not be ready to sleep and you may have a restless night. By observing this simple Happy Hour rule you can make a big difference to your mental and physical wellbeing and feel improvement in your energy and stress levels, as well as happiness.

FOURTEENTH INSTRUCTION:
GET A WATCH!

How many times do you check your phone simply to learn the time? Go "old school' and get a watch. It takes time to get used to, but it's a great way of avoiding unnecessary looks at your phone... and it's a fabulous use of old technology. These practical instructions are the start of your escape from slavery. Follow them with an open mind and, while you may be skeptical to begin with, you will notice an improvement in your relationship with digital devices. You will make your digital experience an altogether more peaceful, relaxed, and fulfilling one and you will develop a keen awareness of how the digital experience affects you. You will also enjoy the feeling of regaining control.

These instructions are a step towards freedom; they are not the final escape. For that will need to go further and purge yourself of all the junk that has trapped you in a cycle of stress, dissatisfaction, and misery. First, we just need to make sure that you are in the right frame of mind to reclaim absolute

FREEDOM

Chapter 15

YOU HAVE NOTHING TO FEAR

IN THIS CHAPTER

•*WHAT YOU'VE ACHIEVED SO FAR* •*SEEING THE REAL PICTURE*
•*CONTROL IS IN YOUR HANDS* •*TIME TO TAKE THE REINS*
•*KNOW YOUR ENEMY* •*REPROGRAM YOUR BRAIN*

In order to achieve the certainty required to conquer your digital addiction painlessly and permanently, you need to remove any lingering fear you might have about your ability to cope after getting free. Approach the subject with an open mind and try to be relaxed, logical, and rational. Then your fears will dissolve.

Easyway works like the combination to open a safe. In order to use the combination successfully you need to know all the numbers and apply them in the correct order. When you began this book you may have found that information frustrating. You were no doubt eager to discover the cure and apply it as quickly as possible.

But you have followed the instructions and have quickly arrived at the stage where you stand on the brink of becoming a Happy Digital User.

You have come a long way toward achieving the state of mind necessary for you to quit and remain free for the rest of your life.

Pat yourself on the back for your achievements and remember how great you will feel when you're free. Allow yourself to feel excited about that. You're releasing yourself from a trap, a prison that has brought you nothing but stress and misery and you're choosing a solution that will enable you to feel happy and in control of your life again.

Perhaps you think you have no reason to congratulate yourself. Perhaps you're still struggling to convince yourself that you can really live happily without your junk apps and that escaping the trap permanently is going to be easy.

It's time to address the fear of success.

SEEING THE REAL PICTURE

Fear of life outside jail can keep the prisoner trapped. The prison feels secure because it's familiar. Even though it's grim inside, the prisoner fears it less than the world outside, which is alien and riddled with uncertainty.

In the context of escaping from digital addiction, we have also established that the fear of success is caused by illusions. These illusions have been put in your brain by many influences, some of which have a vested interest in you remaining hooked.

THE BRAINWASHING

You also fear that escape from the trap will be hard and may leave you feeling more miserable than you do now. You fell into the trap easily and assume that it will be incredibly difficult to get out. But the trap is not a hole in the ground covered with branches and leaves into which you stumbled unwittingly and now find yourself in above your head. Though it may feel like a deep, dark hole, there is no physical effort required to escape. You simply need to make a choice.

It's a simple choice between taking a step backward or a step forward. You can either choose to remain a slave to your phone, to social media, and to online gaming for the rest of your life, feeling more and more trapped and miserable, or you can choose freedom.

Some of our fears are instinctive. For example, the fear of heights, fire, or the sea are instinctive responses that protect us from falling, getting burnt, or drowning. There is nothing instinctive about the fear of getting out of the digital junk trap.

THE FEAR OF COMING OFF DIGITAL JUNK IS CREATED BY USING IT IN THE FIRST PLACE

CONTROL IS IN YOUR HANDS

Once you're free from the digital trap, you'll be amazed at how easy it was to escape. You'll feel much happier and more relaxed. At the moment you may still feel like someone stuck at the bottom of a deep pit, unable to see a way out. But once you get out, you'll realize there was no pit. All you had to do was take a step forward instead of a step back. It's as simple as that.

In order to achieve success you need to remove all doubt. You must understand and accept that your fears of trying to live without your junk apps are based on illusions created by those apps. In reality, you have nothing to fear.

Perhaps you question whether it's possible to know for certain that something will *not* happen, i.e. even if you do manage to get out, how do you know you won't fall into the trap again? What if something happens in your life that makes you feel weak and vulnerable and you get seduced back in? After all, the chances of being struck by a meteorite are infinitesimally small, yet nobody can say with absolute certainty that it will never happen to them.

That is true; however, you have a considerable advantage over potential meteorite victims: If a meteorite is going to hit you, there is nothing you can do about it, whereas only you can make yourself go back to digital junk. But why would you want to when you can see it does absolutely nothing for you?

Addicts only think they need junk for pleasure or comfort because they've been brainwashed. But once you see through the illusion, you can never be fooled by it again. For the rest of your life. There is absolutely no reason to be tempted back into the trap. Remember the rattlesnake: "You knew what I was when you picked me up."

If you still have doubts about this, then there is something you have not understood and you need to go back and reread the relevant chapters, in particular Chapter 7. It is essential that you understand that any desire you had for digital junk was based on an illusion and that once you have seen through the illusion and removed the desire, all you have to do is stop using digital junk and you will never feel the desire to use it again.

YOU WILL NOT MISS IT.

YOU ARE NOT GIVING UP ANYTHING.

If you have followed and understood all the instructions and your desire to quit is as strong as ever, yet you still feel a sense of nervousness, don't worry, that's perfectly normal. It's the excitement of knowing you're about to do something amazing.

TIME TO TAKE THE REINS

You have already come a long way in the process of unraveling the brainwashing that has made you a slave to your devices and putting yourself in the right frame of mind to escape. You have also begun to take the practical forward steps that will see you become a Happy Digital User for the rest of your life.

Your first positive step was choosing to read this book. You had a choice: To bury your head in the sand and continue stumbling further and further into the pit, or to take positive action to resolve the situation.

You made a positive choice. Continue making positive choices and you will succeed.

As we move closer to your moment of escape, there are three very important facts to remember:

1. **DIGITAL JUNK DOES ABSOLUTELY NOTHING FOR YOU AT ALL**

    It is crucial that you understand why this is and accept it, so that you never suffer a feeling of deprivation.

2. **THERE IS NO NEED FOR A "WITHDRAWAL" PERIOD**

    Anyone who quits with Easyway has no need to worry about the withdrawal period. The moment you make your decision to stop using junk is the moment you become free. You don't have to wait for anything to happen—or anything *not* to happen.

3. **THERE IS NO SUCH THING AS "JUST THIS ONCE" OR "THE OCCASIONAL LAPSE"**

    If you are tempted back, it means you never saw through the illusion of pleasure. Escape means seeing through the illusion and removing the temptation altogether. When you achieve that, you have absolutely no desire to ever go back to digital junk. Don't panic if you suddenly become embroiled in some exciting and interesting gossip in one of the groups you remained in. There's nothing wrong with that; it's normal, like spending an evening chatting with friends in a bar, but you wouldn't want to spend every night in the bar with those same friends, would you? Of course not, getting together is special. If you use technology appropriately, then you can pick and choose when to "get together" and when not to.

KNOW YOUR ENEMY

Many addicts suffer from the illusion that they can never get completely free. They convince themselves that their addiction is their friend, their confidence, their support, even part of their identity. And so they fear that if they quit, they will not only lose their closest companion, they will lose a part of themselves.

It's a stark indication of just how severely the brainwashing distorts reality that anyone should come to regard something that is making them miserable as a friend.

When you lose a friend, you grieve. Eventually you come to terms with the loss and life goes on, but you're left with a genuine void in your life that you can never fill. There's nothing you can do about it. You have no choice but to accept the situation and, though it still hurts, you do.

When addicts try to quit by willpower, they feel like they're losing a friend. They know that they're making the right decision to stop, but they still suffer a feeling of sacrifice and, therefore, create a void in their lives. It isn't a genuine void, but they believe it is and so the effect is the same. They feel as if they're mourning for a friend. Yet this false friend isn't even dead. The purveyors of digital junk and all the other addictive substances and behaviors make absolutely sure that these tortured souls are forever subjected to the temptation of forbidden fruit for the rest of their lives.

However, when you rid yourself of a mortal enemy, there is no need to mourn. On the contrary, you can rejoice and celebrate from the start… and you can continue to rejoice and celebrate for the rest of your life.

That's why it's vital to get it clear in your mind that digital junk is not your friend, nor is your addiction to it part of your identity. It never has been. In fact, it's your mortal enemy and by getting rid of it you're sacrificing nothing, just making marvelous, positive gains.

The answer to the question "When will I be free?" is "Whenever you choose to be."

You could spend the next few days, and possibly the rest of your life, continuing to believe that digital junk was an essential part of your life and wondering when you'll stop missing it. If you do that, you will feel miserable, the desire to use it may never leave you; and you'll either end up feeling deprived for the rest of your life or you'll end up going back to using digital junk and feeling even worse.

Alternatively, you can recognize it for the mortal enemy that it really is and take pleasure in cutting it out of your life. Then you need never crave it again and, whenever it enters your mind, you will feel elated that it is no longer ruining your life.

Unlike people who quit with the willpower method, you'll be happy to think about your old enemy and you needn't try to block it from your mind. In fact, it's important that you don't. Trying not to think about something is a sure way of becoming obsessed with it. If you're told not to think about elephants, what's the first thing that comes into your head?

You see!

So you don't need to avoid the subject. On the contrary, you can enjoy thinking about your old enemy and rejoice that it no longer plagues your life.

REPROGRAM YOUR BRAIN

Approaching this process with a relaxed, rational, and open mind helps you to understand the digital trap and the Little Monster that complains when you don't satisfy your cravings. During the first few days after you quit, the Little Monster will keep grumbling away, sending messages to your brain that it wants you to interpret as "I want to check my phone" or "I want to go online to game."

But you now understand the true picture and, instead of complying, or getting into a panic because you can't, pause for a moment. Take a deep breath. There is nothing to fear. There is no pain. The feeling isn't bad. It's what addicts feel after every fix.

In the past your mind interpreted these feelings of withdrawal (the Little Monster) as "I want to check my phone" because it had been brainwashed into believing that doing so would satisfy the empty, restless feeling. But now you understand that far from relieving that feeling, constantly checking your phone actually caused it. So just relax, accept the feeling for what it really is—the death throes of the Little Monster—and remind yourself, "Happy Digital Users don't have this problem. This is what addicts suffer and they suffer it all their addicted lives. Isn't it great! Soon I won't have to suffer it ever again."

At moments like this take a look out of the window, take a look across your office, take a look at the street outside, not as a substitute activity —that would be pointless and perpetuate a feeling of deprivation—but to live in the moment. For just one moment, notice your surroundings: the beauty of a cloud, a tree, a passing dog walker, the architecture of your surroundings. This is life. This is real. Becoming more mindful in these moments is good for your brain, your spirit, and your wellbeing; it's the complete opposite of looking at your phone screen.

THE WITHDRAWAL FEELINGS WILL CEASE TO FEEL LIKE PANGS AND WILL BECOME MOMENTS OF PLEASURE

You might find that, particularly during the first few days, you forget that you've quit. It can happen at any time. You think, "I'll just check my phone", or "I'll go online gaming tonight." Then you remember you don't do that anymore. You have no need or desire for digital junk. But you wonder why the thought entered your head. You were convinced you had reversed the brainwashing. Such times can be crucial in whether you succeed or not. React in the wrong way and they can be disastrous. Doubts can surface and you may start to question your decision to quit and lose faith in yourself.

If these moments happen, don't panic. It's not a bad sign, far from it. Rather than having been obsessed by it, it means you momentarily forgot you'd even got rid of digital junk.

Imagine you had a regular parking space at work, or if you moved home after a few years of being in the same place. If your parking space moved or you moved home, would you be surprised if you occasionally drove into the old parking space or took the route toward your old home over the first few weeks of adjustment? Of course not. Would it mean that you dearly missed your old parking space? Of course not. Would it mean you missed your old home? Well, it might, but normally we're lucky enough to move somewhere better, not worse. These momentary lapses are perfectly normal and do not mean that you're missing the parking space, the old home, or digital junk. These are opportunities for you to live in the moment, and remind yourself how lucky you are to be free.

The mental associations between certain everyday things and digital addiction can linger on after the Little Monster has died and this undermines the attempts of addicts who quit with the willpower method. In their minds they have built up a massive case against digital junk, they've decided to stop using it, they've managed to go for however long without using it and yet on certain occasions a voice keeps saying, "Check your phone." Maybe they see someone else checking their phone, or they hear another phone bleep and it reminds them of their own. It triggers the old desire. They haven't killed the Big Monster, and so they still think of digital junk as a pleasure or comfort.

When you quit with Easyway, you no longer suffer the illusion that you're being deprived, yet it's still imperative that you remember the "parking space" and "old home" analogy for the occasions when these potential triggers occur. If you momentarily forget that you no longer use digital junk, that isn't a bad sign. It's a very good one. It's proof that your life is returning to the happy state you were in before you got hooked, when digital devices didn't dominate your whole existence.

Expecting these moments to happen and being prepared for them means you won't be caught off guard. You'll be wearing a suit of impregnable armor. You know you've made the correct decision and nobody will be able to make you doubt it. That way, instead of being the cause of your downfall, these moments can give you strength, security, and immense pleasure, reminding you just how wonderful it is to be

FREE!

Chapter 16

PREPARING TO QUIT

IN THIS CHAPTER

•*CHECK YOUR STATE OF MIND* •*PICKING YOUR MOMENT TO QUIT*
•*GETTING IN THE ZONE* •*START LETTING GO*
•*FORGOTTEN SOMETHING?*

You have come a long way since you first picked up this book. If you compare your mindset then with your mindset now, you should find that you now have a much clearer understanding of the psychological trap you were in and you should feel ready to walk out of that trap. Rather than any feeling of impending doom and gloom, you should have a sense of excitement and anticipation. In fact, you've probably noticed some changes in your behavior and attitudes already. The door is open. Are you ready to walk out?

Don't be surprised if you feel nervous. You are about to walk free from an evil trap that has kept you enslaved and taken over your life, making you stressed, exhausted, and miserable. It's caused almost irreparable damage to some of your closest relationships and probably helped cement countless pointless and meaningless relationships in the meantime. You're about to reclaim control and banish those feelings of slavery and powerlessness for good. If you've collected some negative influences in the form of an acquired "Cyber Family," then jettison them. There's nothing wrong with maintaining friendships with people you've met online; just select the meaningful ones to retain and stay in touch with.

Very soon you will be able to call yourself a Happy Digital User. Take pride in your achievement; there isn't a digital addict in the world who wouldn't wish they could do what you are about to do. You are about to discover the unbridled joy of having nothing to hide, having time for the people you love and the things you love to do. While you were in the trap you lost sight of the genuine pleasures in life. You are about to get that huge part of your life back.

Conquering addiction with Easyway is a thrill, but in practical terms it's more like preparing for a pleasant walk in the country than something potentially terrifying like a skydive. The fact is simply that you're about to walk out of a prison cell. If this isn't how you feel, if you have doubts about what you are about to do, it means that you haven't understood a key part of the method and you need to go back and reread the relevant chapters until you do. To help you, take a look at the code word, EASYWAY, on the next page. This serves as both a reminder and a checklist. Go through each item and ask yourself:

- Do I understand it?

- Do I agree with it?

- Am I following it?

If you have any doubts, reread the relevant chapters as listed.

E   ESCAPE

It's easy when you use the right method. You're not making a sacrifice, you're choosing a better way of life, so you have nothing to fear. (See Chapter 5)

A   ADDICTION

Acknowledge that you have become addicted and you can stop blaming yourself. Breaking the vicious circle of "need," fix, and withdrawal means reversing the brainwashing and seeing through the illusions. Remember, the circle started as fix, withdrawal, "need." (See Chapter 6)

S   SACRIFICE

There is no sacrifice. Digital junk does nothing for you whatsoever, so you're not giving anything up. The only things dying are the two monsters that have kept you enslaved. (See Chapters 7 and 8)

Y   YOU

There is nothing wrong with you. You don't become an addict because you have an addictive personality; you become an addict because you take an addictive substance or begin an addictive behavior and engage in activities that are specifically designed to addict you. You are your own jailer and you have the power to escape. (See Chapter 10)

W   WILLPOWER AND WITHDRAWAL

Relying on willpower makes it harder to quit and highly unlikely that you will stay free. You only need willpower if you have a clash of wills. Withdrawal is what addicts feel after every fix. When you quit with Easyway, there is no unpleasant withdrawal feeling or withdrawal period. You become free the moment you change your mindset and remove any desire for a fix. Enjoy the feeling of being free—it is the Little Monster dying. (See Chapters 9 and 15)

A   ACTION

Don't wait for a miracle. It's time to act. Follow the instructions and you cannot fail to get free. Keep an open mind, let common sense and logic replace illusion and very quickly you will find your relationship with digital devices has changed completely. (See Chapter 1)

Y   YES!

Enjoy the thrill of being free. Never doubt your decision. You are not depriving yourself or losing a friend; you are destroying a mortal enemy and getting your life back. (See Chapter 15)

Stopping with Easyway is not difficult. All you need to do is follow the instructions and you will succeed. Any resistance you feel is brought about by the brainwashing, which tells you that quitting cannot be easy. Unravel the brainwashing and the resistance evaporates.

You have already done the work that was necessary to put yourself in the right frame of mind. Your training and preparation are almost complete. You are fully equipped to succeed at something that other ex-addicts regard as the most important and significant achievement of their lives. If you feel like a dog straining at the leash, eager to get on with it, that's great, but you still need to concentrate carefully on the rest of the book.

Very soon you will be ridding yourself of the junk that has kept you enslaved and making a vow never to fall into that trap again. You may be wondering when would be the ideal moment.

PICKING YOUR MOMENT TO QUIT

There are two typical occasions that tend to trigger attempts to quit any sort of addiction. One is a traumatic event, such as a health scare or a tragedy that strikes someone close; the other is a "special" day, such as a birthday or New Year's Day.

Let's call these "meaningless days," because they actually have no bearing whatsoever on your digital addiction, other than providing a target date for you to begin your attempt to stop. That would be perfectly OK if it helped, but meaningless days actually cause more harm than good.

New Year's Day is the most popular of all meaningless days, being a clear marker of the end of one period and the beginning of another. People who try to quit on New Year's Day also happen to have the lowest success rate. The Christmas holidays are a time when we binge more than usual and by New Year's Eve we're just about ready for a break.

As part of the whole New Year celebration, you make a resolution to quit digital addiction: "This year will see the new me."

After a few days' abstinence, you've recovered your balance and are feeling cleansed. The self-loathing that strengthened your will to quit subsides, but the Little Monster is screaming for its fix. If you're relying on willpower and don't understand that feeding the Little Monster will only make it worse, you give in and allow yourself "just the one." The Little Monster is quietened for a while and you think to yourself, "I really need this. It makes me feel good."

JUST THE ONE POINTLESS JUNK GAME IS ALL IT TAKES TO DRAG YOU BACK INTO THE TRAP

Meaningless days only encourage you to go through the damaging cycle of half-hearted attempts to quit with willpower, bringing on the feeling of deprivation, followed by the sense of failure that reinforces the belief that stopping is very difficult. Addicts spend their lives looking for excuses to put off "the dreaded day." Meaningless days provide the perfect excuse to say, "I will quit, just not today."

Then there are the days when something shakes your world and you respond by saying it's time to sort yourself out. But these stressful times are also when the pull of your addiction becomes strongest because you regard it as a form of comfort. This is another ingenuity of the trap:

NO MATTER WHICH DAY YOU CHOOSE TO QUIT, IT ALWAYS SEEMS TO BE WRONG

Some addicts choose their annual vacation to quit, thinking that they'll be able to cope better away from the everyday stresses of work and home life and the usual temptations to look at devices. This approach might work for a while, but it leaves a lingering doubt: "OK, I've coped so far, but what will happen when I go back to work and the stress builds up again?"

When you quit with Easyway, it doesn't matter when you quit— the prevailing conditions have no bearing on your chances of success because you are not relying on willpower. You can (and should) go out and handle your usual challenges straightaway, so that you can prove to yourself from the start that even at times when you feared you would find it hard to cope without your regular fix of digital junk, you're still happy to be free.

When you're stuck in the trap trying to will yourself free, it's like trying to open a door by pushing on the side where the hinges are, rather than the side that swings open. If you had been pushing fruitlessly on the wrong side of a door and then you discovered you could open it easily by pushing on the other side, would you wait until New Year's Day? Or until your birthday? Or until you went on vacation? Or would you make your move there and then?

FIFTEENTH INSTRUCTION:
DON'T WAIT FOR THE RIGHT TIME TO QUIT. DO IT NOW!

GETTING IN THE ZONE

You have everything you need to quit. Like an athlete on the blocks at the start of the Olympic 100 meters final, you are in peak condition to make the greatest achievement of your life RIGHT NOW! Now you just need to get yourself in the zone.

Focus on everything you have to gain: A life free from slavery and stress and a reconnection with genuine pleasure. No more feeling like a lab rat, being manipulated and controlled by unseen digital designers; no more closing yourself off from the people around you; no more missing out on real social interaction; no more sneak peeks at your phone and skulking off to the restrooms to check your notifications; no more running to the front door only to find brown envelopes.

In place of all that slavery you can look forward to living in the light, with your head held high, enjoying open, honest relationships with the people around you, feeling healthy, energetic, able to relax, in control of how you spend your time, and finding real enjoyment in the genuine pleasures that you enjoyed before you became hooked. The great news is that with greater mindfulness, greater connections with those closest to you, you're all set to enjoy heightened senses of pleasure: The joy of touching, meeting, being together with someone, or simply doing nothing at all.

"SORRY, DID YOU SAY 'DOING NOTHING?'"

Yes, don't be afraid of nothingness or moments of boredom. When you look back on the last few days, wouldn't you give your eye teeth for a few moments to yourself of doing absolutely nothing?

Boredom, not really having anything to do, is an important part of your wellbeing, yet we consider it an enemy.

It's like our attitude to hunger, which has been manipulated by sugar addiction and junk food. We do everything to avoid the feeling of hunger, but hunger itself is a great feeling. Clearly not if you're suffering from hardship or malnutrition, but in everyday life it's something we're infinitely better off feeling, rather than never feeling.

Fear of hunger or unfamiliarity with hunger causes us to overeat and feel bloated and gorged all the time. As soon as we sense the remotest feeling of hunger, we stuff something into our mouths, normally of zero nutritional value and full of refined sugar or processed carbohydrates. We don't realize that by gorging all the time we completely lose touch with one of the gentlest, most pleasing aspects of our existence—feeling slightly hungry and enjoying letting that feeling develop until our appetite naturally indicates that it's time to eat. We then eat sensibly, in a planned way, and rather than become bloated on junk food we become genuinely satisfied without that awful, heavy, bloated feeling.

The same goes for boredom; silence, solitude, and quiet nothingness can be a beautifully calming thing. Make friends with those wonderful moments rather than run away from them. Taking a moment to recognize your surroundings, looking at a picture on the wall for the first time in years, or taking a look outside in the garden is the complete opposite of digital junk.

If you are clear on all the points in the EASYWAY checklist and have followed and understood all the instructions, then you should be feeling the relief of knowing that the illusions that surround digital addiction (the Big Monster) have been destroyed. If you haven't had an amazing moment of revelation, don't worry. For most people this occurs a few weeks down the line when they suddenly realize that they're completely free.

Either way, you are ready to quit.

START LETTING GO

The following instructions will help you to regain your independence from your devices.

SIXTEENTH INSTRUCTION:
BOUNDARY YOUR DEVICES

The Thirteenth Instruction in Chapter 14 was to observe an hour of screen-free time before going to sleep at night. This was to allow your brain to unwind and to avoid the confusion caused by the blue light emitted from screens. Now you need to go further and banish your devices from the bedroom altogether.

If you sleep with your phone by the bed, your first instinct when you wake up is to reach for it and start checking your notifications. Even when you've killed the Big Monster, this behavior can persist, especially if you're sleepy and not fully aware of what you're doing. It doesn't mean you're still hooked, or that you've fallen back into the trap. It can be a reflex programed by routine rather than any desire to look at your phone, but it can still be damaging to your belief that you have killed the Big Monster. After all, what kind of a start to the day is to be inundated with countless messages, emails, and alerts that even healthy digital use generates? By sleeping with your phone (and other devices) in another room, you take this danger out of the equation.

If you normally use your phone as an alarm clock, you have a choice: Either continue to use your phone to wake you up, but keep it in another room so you really do have to get up when it goes off; or buy an alarm clock. Rediscovering old technology can be fun and liberating.

The important thing is to make sure that every time you check your devices it's a conscious decision and you're not using them mindlessly.

Putting your phone out of reach means you will only use it when you really need to. So as well as banishing it from the bedroom, make a rule to keep it off the table in restaurants and put it out of reach in the car. Even when you've killed the Big Monster, you can convince yourself that a phone call is important enough to take priority over everything else you're doing—including keeping your eyes on the road. Be very clear: it is not.

*EVERYTHING THAT COMES IN VIA DIGITAL DEVICES
CAN WAIT. HITTING THE BRAKE WHEN THE DRIVER
IN FRONT APPLIES THEIRS CANNOT*

SEVENTEENTH INSTRUCTION:
PURGE YOUR EMAIL

What percentage of the emails that come into your inbox are actually important and what percentage are spam? Unless you're extremely lucky, the vast majority will be spam. Spam is incredibly annoying and can be very stressful. Just seeing that you've received tens or hundreds of emails since you last logged on will cause anxiety, regardless of whether or not most of them can be deleted immediately without opening them.

You can put a dent in the barrage of spam emails by setting time aside to unsubscribe, where that option is available. You can also set up filters to block emails from specific senders. Digital addicts find it very hard to find the time to carry out this sort of maintenance because it gets in the way of their digital fix, but you no longer have that distraction. Half a day spent deleting, unsubscribing, and blocking unwanted emails could make a big difference to your email experience. Remember to renew that purge every now and then... it's amazing how often they build up again.

You simply don't need to know about dozens and dozens of theater shows over three years just so that you can be notified of one, yes one, that you might actually be interested in.

Get rid of all those theater and movie alerts; you rarely read them, let alone respond to them anyway.

If you continue to receive unwanted emails, consider changing your email address. This may be a nuisance in the short term, but in the long term it could save you a lot of stress. By starting afresh with a brand new email address, you can protect that email from unwanted interventions by using it solely for emailing—not as your user name for logging on to websites and apps, which can then be hacked and used to bombard you with spam.

If you do ever need to log into free wifi in coffee bars, restaurants, airports, etc., although you'll be far less likely to do so now you're no longer a digital junkie, always do so with a fake email address. No, believe it or not, you don't need to give your real email address to access free wifi in these places.

Once you have purged your inbox and stemmed the flood of spam, set about reducing the so-called important emails in your inbox. If you have anything that's more than three months old, delete it. That's right, delete it.

Hoarding old emails is like hoarding boxes of junk in your attic. All it does is take up space and occupy your mind. If you consider some of these emails important, then file them accordingly. Don't leave them clogging up your inbox. As long as it's there, you will have to think about it from time to time. "Do I need to keep this?" "Should I be doing something with this?" Get rid of it once and for all and free your mind from these unnecessary considerations.

If there are recent emails that you do need to keep for reference, don't leave them in your inbox. Move them to a new mailbox labeled according to the sender or issue that they pertain to. This filing system makes it easier to find the emails you need.

Other than for those whose working lives rely on constant email bombardment, your new regime of logging on three times a day, rather than leaving your email on full time, purging, and sorting your emails will give you a much more efficient, enjoyable email experience with far less junk to wade through every time you log on.

For heavy email users, follow the advice provided earlier. Do what you can to delegate emails, resist having unnecessary emails forwarded to you, and GET HELP. It's amazing how many businesses fall into decline because the business owner—this applies just as much to a sole trader as to a bigger business—skimps on having someone to field telephone calls and emails. Use tech to your advantage; you can now have virtual assistants that cost a fraction of a full-time employee. You might even become so much more productive and profitable that you DO need to employ someone full time. What an absolute joy that is: building something for the future rather than simply making do, or "surviving."

EIGHTEENTH INSTRUCTION:
GO NAKED

Sleeping with your phone in another room is the start of the process of separation that proves you have broken free of your addiction to digital devices. The next step is to spend your waking hours apart.

Start in the evenings. Try turning your phone off two hours before going to bed, then three hours, then four. Turn it off and put it somewhere out of sight.

Get used to going about your everyday routines—cooking, cleaning, relaxing, socializing—without having your phone constantly to hand. Enjoy the freedom, the space. Did you watch that first movie with your phone turned off yet? What a beautiful, relaxing, joyous release. As long as you chose the right movie of course.

At first you might feel naked without your phone. That's not a bad thing. Being naked can be a very liberating feeling. In this case, you're not really naked so you don't have to worry about the consequences of being seen! Focus on the feeling of being without your phone and rejoice in the knowledge that you have come this far. You have learned how to survive on your own again, without your perceived crutch.

Over the weekend, try going without your phone for a whole day, then both days. It's not so long ago that this was the norm. Apart from the home phone, no one had any form of telecommunication to keep them in the loop. It's slightly scary to even think about it, but why not do it just once to see what it feels like?

If you need to call people, buy a dumb phone and use it purely for making and receiving calls.

The important thing is to get back to the feeling of independence that you had before you became reliant on your phone and other devices to keep you in contact.

The same applies to other devices like fitness trackers. Do you really need a digital device to tell you whether you're in good shape or not? Isn't that something you feel for yourself? Tracking your steps and other biometrics can quickly become obsessive and affects your natural, in-built self-monitoring program. You can feel fit and healthy, but if the tracker shows you're falling short of your targets, you won't be happy.

Conversely, you can know inside that you are overweight, but if the tracker shows that you're hitting your exercise targets, you will kid yourself that you're in good shape.

Put the trackers and other digital devices aside and reconnect with your sense. You will feel empowered and the results will be more in line with what you really want.

If you really want to stay connected to your tracker, then be sure to turn off all the notifications, reminders, and alerts. Other than one that might remind you to move around to avoid DVT and another to notify you when you've completed your target steps, that should be it. Can you think of anything else a tracker really needs to do for you?

FORGOTTEN SOMETHING?

These instructions all go against the barrage of brainwashing from Big Digital. They all challenge the widely held belief that digital junk is essential for a happy life in the digital age. The brainwashing is incredibly powerful and once it has you hooked it increases, via the junk apps, junk games, and junk social media platforms that it pushes on you.

But when you do stop to challenge the brainwashing and put it to the test, the cracks appear and the illusions are dispelled. By following these instructions, you will see for yourself that it is perfectly possible to live a happy life without being a slave to digital. Not only that, it is much easier to lead a happy life without being a digital addict.

There is one instruction you might have expected but have not yet been given: the instruction to get rid of your junk apps. Just think of it as saving the best bit till last.

Chapter 17

ESCAPE!

The time has come to complete your escape. Your mind is right, you have all the information you need and a life of happy, healthy digital use awaits you. All that's left to do is mark the occasion with a ritual that will stay with you forever.

At the start of this journey you were asked not to change the way you use digital devices, other than a few tips and exercises along the way. It was important that you weren't sidetracked into trying to limit your use with willpower, which would have caused too many distractions for you to follow and understand the instructions and logical arguments along the way. Since Chapter 14 you should have begun to follow the practical instructions for unhooking from your devices. You are now ready to take your final step to freedom and make the change from being a digital addict to being a Happy Digital User.

NINETEENTH INSTRUCTION: DELETE YOUR JUNK APPS, BAIL OUT OF YOUR JUNK GROUPS, AND PURGE YOUR FRIENDS' LISTS, SO THAT YOU ONLY INCLUDE THOSE YOU REALLY NEED, AND REALLY WANT, TO BE IN TOUCH WITH

By now you should have sorted the apps on your devices into "Useful" and "Junk."

Remember, junk apps are the apps that promise pleasure or comfort, but more often than not deliver frustration, insecurity, anxiety, and disappointment. Isn't that an apt description of much of social media? By using your sense—and really paying attention to how you feel when you use these apps—you will be able to make that decision for yourself. You don't have to withdraw from social media, unless you really want to, but be sure to only retain the elements that inform, inspire, and help you stay in touch with loved ones. REAL loved ones. That means close friends and family.

It's time to dump the junk apps out of your life once and for all. Deleting an app is very simple.

Take great delight in deleting those apps and watching them disappear.

THE FINAL MESSAGE

There'll be some friends who perhaps you haven't retained on your social media friends' lists but don't wish to offend and don't wish to lose touch with. Reach out to them. Explain that you've dramatically trimmed your social media use because it was causing you real problems, but that you don't want to lose touch with them, so they know to contact you directly if they ever want to meet or simply have a chat on the phone.

The same goes for groups. If you feel you want to stay in touch, and not alienate yourself from a group, by all means drop the group a line with your Final Message.

It's a simple statement, explaining that you are trimming down your social media use because it was causing you a problem. Feel free to, but don't feel obliged to, reveal that you had become a digital addict but have cured yourself by using Allen Carr's Easyway. If you want, you can go into detail about how your addiction made you feel and the effect it had on your mental and physical health.

You might add that you are looking forward to seeing everybody in the real world.

There is a simple, practical purpose to this message. You want your friends to know that you're no longer participating, so they know to use other channels when they want to get hold of you. But there is a powerful psychological purpose to the final message too.

In a world where social media plays such a controlling part in so many people's lives, a message announcing that you are quitting is immensely powerful.

It will make your friends and contacts think twice about their own use of digital junk. Don't be surprised to see others following your lead in the weeks and months to come.

The Final Message is also empowering for you. It's like drawing a wonderful line under your addiction.

You don't need to be a drama queen about it—it's just a communication of your new digital attitude.

Social media is all about conforming to views and behaviors that you think will make you accepted, i.e. following the flock. The final message bucks that trend with a vengeance. It is your way of saying to yourself, "I don't need to conform. I am happy being myself and following my own rules."

The Final Message also gives you the immense satisfaction of turning the tables on the digital designers who have put so much research and intellect into making you an addict. Now you're using their tools to complete your escape!

THE RITUAL

When you've written your Final Message, take a deep breath, savor the moment, and press "send." As you click the button, remind yourself that:

- Digital junk gives you no genuine pleasure or comfort.

- Digital junk doesn't relieve stress, anxiety, or isolation; it causes them.

- The only reason you ever thought it did was because you were brainwashed.

Now close your eyes and make a solemn vow that you will never be seduced into using digital junk again. **EVER!** Be certain about the decision you're making, embrace the moment, and greet it with a sense of triumph.

The most important purpose of the ritual is this: The thing that makes it difficult to quit is not the physical aggravation of withdrawal feelings; it's the doubt, the uncertainty, the waiting to become cured. With Easyway, you become a Happy Digital User the moment you complete the ritual of the Final Message. It is not a gradual process; it is a moment in time. It's important to know when that moment is and remember it forever.

If your final message is for buddies that you online game with, face up to it: These are either people who you only have one thing in common with (highly addictive online gaming) or friends that you also spend time with while engaging in highly additive online gaming.

If it's the former, you need to ask yourself a question: Are they friends? Real friends? If so, then you can arrange to hook up with them from time to time (if you live on the same continent as them, obviously). If they're not, wish them good luck and depart; you'll just be a weird MIA figure in their fantasy world of toy soldiers, secret zombie missions, and tribal battles, or whatever it is you've spent time with them doing.

If they're real-life friends, who you normally see at social gatherings from time to time or frequently, then there's no loss. You'll continue to see them.

Think about how those games have made you feel. The feelings were phoney. Did they make you feel strong, powerful, speedy? Did you feel like an international super-assassin? Disown those phoney feelings. You're Kevin from Sales, not "The Dark Destroyer" or whatever your online name is.

Make Kevin from Sales a real hero. Do stuff. Enjoy REAL life. Wake up and smell the coffee. Spend time and energy with your loved ones. Take real pleasure in real-world success. No matter how small that success might be, IT'S REAL. People love Kevin, and more importantly Kevin loves Kevin.

And holding your partner's hand, or calling your parents, or visiting your brother or sister, or just chilling out watching a movie is valid. Scoring the winning run for the Sunday baseball team or even losing 7–0 will provide a hundred times the pleasure, real pleasure, than scoring for Barcelona vs. Real Madrid on *FIFA 20*. If you're getting on a bit, try taking a walk or just watch a real baseball match. Taking your son, daughter, nephew or niece to the park to play on the swings makes you a REAL hero, not the phoney fakery of slaying a zombie warlord and his psycho-killer army. Ditch it. Forever. And say goodbye to those you leave behind.

Have others left before you? Probably. Will you be the last to bail out? Definitely not.

If you don't have a job or a partner and feel lonely and isolated, get out there in the real world and lap it all up. Take small steps, but out there is a wonderful world of real experiences, real fun, real enjoyment. Even the occasional pain is real and something that will make you stronger rather than just selecting "Ready Player One."

Whether your digital addiction has been via social media, junk apps, smartphone games, or full-on online gaming (or all of the above), it's completely normal to feel nervous at this stage. If you have a few butterflies in your stomach, that's fine; it's no threat to your chances of success.

Rather like making a parachute jump, your nervousness will quickly turn to exhilaration as you realize that everything you've learned and prepared for is working exactly how you've been told it would. The feeling is one of incredible freedom and elation. Imagine the thrill of simply walking out of a dark, damp, cold dungeon and walking into a world of sunshine, fresh air, happiness, and freedom.

It's far from a parachute jump. You are simply standing by a prison door, ready to push gently on the correct side of the door. You have all the knowledge and understanding you need to make this the best, easiest experience of your life. The nerves are perfectly natural and normal. Soon you will be free.

DON'T WAIT FOR THE OUTCOME

Your escape is complete from the moment you finish your Final Message and make your vow. The ritual marks the start of your new life, the breaking of the cycle of addiction. And that's it. There is nothing to wait for. You're now a Happy Digital User.

Enjoy your victory. This is one of the greatest achievements of your life. It's important that this moment is firmly implanted in your mind.

Right now you are fired up with all the powerful reasons to stop, but be aware that in a few days your resolution will fade and you will relax into a normal life. As the days, weeks, and years go by, your memory of how miserable digital addiction once made you feel will dim.

So fix those thoughts in your mind now while they are vivid. Write them down, describe how low digital junk dragged you down and affected your life, and every now and then take a look at that description —not to scare yourself from going back but to smile and reflect on the joy of your freedom.

CONGRATULATIONS!

YOU'RE FREE!

Chapter 18

ENJOYING LIFE FREE FROM DIGITAL ADDICTION

IN THIS CHAPTER
•WITHDRAWAL •WHEN THINGS GO WRONG IN LIFE
•START LIVING NOW

You have done a wonderful thing! You are now ready to get on and enjoy the marvelous pleasures of life free from the worthless constraints of digital addiction. Just make sure you remain aware that there will be moments when you need to remind yourself of the key points that have helped you reach this happy day.

For a few days after you've quit, you may still feel the Little Monster crying as it goes through its death throes. There is no need to fear this feeling, but you shouldn't try to ignore it either. Now that you're aware of what it is, you can congratulate yourself on taking control and enjoy feeling the Little Monster die.

This is the withdrawal feeling that is made out to be such an ordeal for addicts, especially by people who have tried to quit with the willpower method. For them it often is an ordeal, because they respond to the Little Monster's death throes by thinking, "I need to check my phone." This triggers a mental struggle, which causes the physical symptoms associated with withdrawal.

WITHOUT THAT MENTAL STRUGGLE, THE WITHDRAWAL PERIOD IS NO PROBLEM

Living with the death throes is no harder than living with a slight itch and they only last for a few days. They only become a problem if you start to worry about them or interpret them as a need or desire for digital junk. If you do feel them, picture the Little Monster searching around the desert for food and you having control of the supply. All you have to do is keep the supply line closed. It's as easy as that.

Instead of thinking, "I want to check my phone, but I'm not allowed to," think, "This is the Little Monster demanding its fix. This is what digital addicts suffer throughout their addicted lives. Happy Digital Users don't suffer this feeling. Isn't it great! I'm a Happy Digital User and so I'll soon be free of it forever." Think this way and those withdrawal pangs will become moments of pleasure. Remember the parking space and new home analogy. Thinking about something doesn't mean you have a problem with it. Enjoy those moments of realizing that you're free.

Remember to pay attention to what your senses are telling you. Focus on the feeling and allow yourself to become aware that there is no physical pain—the only discomfort you might be feeling is not because you've stopped using digital junk, but because you started using it in the first place.

Also be clear that if you were to try to stop the feeling by using digital junk again, far from relieving that discomfort, you would insure that you suffered it for the rest of your life.

Take pleasure in starving that Little Monster. Revel in its death throes. Feel no guilt about rejoicing in its death.

WHEN THINGS GO WRONG

Of course, there will be days when you find it hard to see the joy in life. That's perfectly normal. Everybody has days when everything that can go wrong does go wrong. It has nothing to do with the fact that you've quit being addicted to digital junk. In fact, when you stop using junk you find that the bad days don't come around so frequently.

For people who quit with the willpower method, bad days can be the trigger that leads them back to digital addiction. Because they don't understand about the brainwashing, even long after the physical withdrawal has ended, they will interpret normal feelings of stress or irritability or loneliness as a need or desire for digital junk. They won't want to cave in, though, because they have made a huge effort to quit, so they will feel deprived and that will make the misery worse.

Sooner or later their willpower will give out and they will "treat themselves." They'll tell themselves it's "just the once," but very soon they will find themselves hooked again. If their willpower doesn't give out, they will spend the rest of their lives enduring the agony of wondering when the sense of deprivation will end.

You might well find that when you have bad days, the thought of burying your head in your phone enters your mind. Don't worry about it and don't try to push it to one side either. Remember the elephants! You cannot tell your brain to *not* think about something. If you try to *not* think about using digital junk, you'll think about it even more and will get frustrated and miserable.

But that doesn't mean you're putting yourself at risk. When you have no desire to use junk, you can think about it all you like. What's more, you can remind yourself of the marvelous truth. Whereas someone who tries to quit with willpower will think, "I mustn't check my phone," or, "I thought I'd beaten this craving," you'll be thinking, "Great! I'm a Happy Digital User! I'm free!"

More importantly, rather than reaching out to imagined friends online, or groups of sycophantic "Cyber Family" who will simply tell you in messages of no more than 280 characters whatever it is they think you want to hear, you're more likely to reach out to a real friend, or strike up a conversation with a real person, and have a real talk, and feel infinitely better as a result.

The sixth instruction was never to doubt or question your decision to quit. This is essential. If you allow doubt to creep in, you will allow the Big Monster back in and soon you'll be back in the trap.

Prepare yourself for the bad days and have your mindset ready. Protect yourself against getting caught out by them. Be ready for feelings of stress, irritability, sadness, loneliness, boredom, disappointment, or apathy and remind yourself that you are better equipped to handle them now than you were as a digital addict. Using digital junk would only make them worse.

Be absolutely clear that there is no such thing as "just the once." If the thought of having "just the one" ever enters your mind, replace it with the thought, "Yippee! I'm a Happy Digital User! I have set myself free from that life of misery." You'll experience real joy rather than worry!

START LIVING NOW

The wonderful thing about overcoming digital addiction with Easyway is not just that it takes away the struggle, but that you don't have to wait for the Little Monster to die before you start enjoying life again. Freedom begins the moment you send your Final Message and make your vow. Throw caution to the wind… pop out to the shops without your phone. Just for an hour. See what happens. Nothing bad will happen.

IT'S TIME TO GET ON WITH LIFE

On the one hand you're freeing yourself from slavery; on the other you're about to rediscover life's genuine pleasures. It's a win-win.

Digital addiction, like all addictions, takes away the ability to enjoy the things that you used to enjoy: reading books, watching entertainment, social occasions, exercise, relationships… Because you regarded your little fix as the only thing that can relieve your craving, nothing else gives you satisfaction. Now that you are a Happy Digital User, you have all these pleasures to get excited about again.

You will find that situations you have come to regard as unstimulating or even irritating become enjoyable again: things like spending time with your loved ones, going for walks, seeing friends, a little bit of boredom. The chance to look out of the window. Work too will become more enjoyable, as you find you are better able to concentrate, to think creatively, and to handle stress.

Most importantly, you'll be able to make the most of the marvelous benefits of the Digital Revolution. Becoming free of digital addiction does not mean never using digital devices again. It's what you use those devices for that matters. Once you've removed the desire to use them for junk, you have the power and control to do whatever you want on your devices, shielded from any attempts to drag you back into the trap.

The Digital Revolution has evolved very quickly and we have been overwhelmed by the multitude of devices and apps presented to us. We have forgotten that we have a choice and have blindly followed the flock. As time goes by, however, more and more people are beginning to question the benefits of all this technology. "How much of this do I really need?"

Consumers are noticing the increased stress, the dwindling free time, and are looking for ways to curb their dependence on digital devices.

Even the founders and top executives of these global corporations have taken measures to limit their exposure to the digital addictive nature of their products; more than that, they keep their kids safe from them for as long as they can too.

And they've actually "gone public" about having done so!

You have just gotten ahead of the game. As the Digital Revolution continues to evolve, with ingenious new apps coming and going, the smart users will be those who refuse to be swamped by brainwashing but use their own senses to pick and choose the apps that are genuinely beneficial to their lives. Your instincts have led you to read this book.

You are now a "smart user." By sharing the good news about your escape, you will help other digital addicts to become smart users too.

That's something to be proud of.

YOU'RE SMART PHONE, NOT DUMB PHONE!

Chapter 19

USEFUL REMINDERS

From time to time you may find it useful to return to this book to remind yourself of some of the things you have learned. Here is a summary of the key points, together with a reminder of the instructions. Follow these and you will remain a Happy Digital User for the rest of your life.

If you have turned straight to this page, hoping to find a shortcut to the solution to your digital addiction, sorry, but that won't work. You need to start from the beginning and read all the way through the book in order. Once you've done that, the following reminders will make perfect sense.

- Don't wait for anything. You are free from the moment you unravel the brainwashing and kill the Big Monster and you can get on with life as a Happy Digital User as soon as you complete the ritual of the Final Message. You have cut off the supply to the Little Monster and unlocked the door of your prison.

- Accept that there will always be good days and bad days, but remember that you will be stronger both physically and mentally without your addiction, so you will enjoy the good times more and handle the bad times better.

• Be aware that a very important change is happening in your life. Like all major changes, including those for the better, it can take time for your mind and body to adjust. Don't worry if you feel different or disorientated for a few days. It's all part of the wonderful achievement of getting free.

• Remember you've stopped bingeing on digital junk: You haven't stopped living. You can now start enjoying life to the full and rediscover all the genuine pleasures and comforts life has to offer.

• There is no need to avoid other digital addicts just as long as you communicate in real life rather than online. Go out and enjoy social occasions and show yourself you can handle situations without feeling tempted to reach for your phone from the start. Your example will encourage others to follow your lead. But, please don't turn into a "reformed addict." You know the type. They bang on at people who are in the grip of an addiction that they've escaped from, making other people feel inadequate and awkward. Most of these types survive purely on willpower and have to reassure themselves they've made the correct decision by making other people feel inadequate. You won't feel compelled to be one of those, but you'll be perfectly happy to point someone in the right direction who notices a change in you and asks about how you became free.

• Certainly don't envy digital addicts. When they're behaving like you used to, buried in their phones; remember that you're not being deprived; they are. If they knew about you, they would envy you and wish they too could be like you: **FREE.**

• Never doubt or question your decision to stop—you know it's right. If the thought enters your head that life will be less enjoyable without digital junk, remember how miserable it felt to be in the grip of the Big Monster. If you allow temptation to creep in, you'll put yourself in an impossible position: miserable if you don't and even more miserable if you do.

• Make sure right from the start that if the thought of "just one binge" enters your mind, you respond with the thought, "YES! I no longer have any desire to do that. I'm a Happy Digital User." Think about the car parking space and new home analogy. These become wonderful moments to enjoy freedom.

• Don't try *not* to think about using digital junk. It's impossible to make your brain not think about something. You will only make yourself frustrated and miserable. It's easy to think about junk without feeling miserable: instead of thinking, "I mustn't use it," or, "When will the craving stop?" think, "Great! I'm a Happy Digital User. Fantastic! I'm free!"

• Share your achievement. Come clean with everyone who has been affected by your addiction and be proud to talk about your decision to quit to anyone who asks. Other addicts will be truly inspired by you.

YOU HAVE DONE A WONDERFUL THING. ENJOY YOUR FREEDOM!

THE INSTRUCTIONS

1. FOLLOW ALL THE INSTRUCTIONS

2. DO NOT ALLOW YOUR DEVICES TO INTERUPT YOU WHILE YOU ARE READING THIS BOOK

3. SEE YOUR SITUATION FOR WHAT IT REALLY IS

4. OPEN YOUR MIND

5. BEGIN WITH A FEELING OF ELATION

6. NEVER DOUBT YOUR DECISION TO QUIT

7. IGNORE ALL ADVICE AND INFLUENCES THAT CONFLICT WITH EASYWAY

8. GET REAL ABOUT YOUR DIGITAL USE

9. PAUSE AND PAY ATTENTION

10. PERSONALIZE IT FOR PROGRESS

11. IGNORE EMAIL

12. DISABLE PUSH NOTIFICATIONS

13. OBSERVE HAPPY HOUR EVERY DAY

14. GET A WATCH!

15. DON'T WAIT FOR THE RIGHT TIME TO QUIT. DO IT NOW!

16. BOUNDARY YOUR DEVICES

17. PURGE YOUR EMAIL

18. GO NAKED

19. DELETE YOUR JUNK APPS

20. DON'T THINK: "I MUSTN'T" OR "I CAN'T", THINK:

"YIPPEE! – I'M FREE"

ALLEN CARR'S EASYWAY CENTERS

The following list indicates the countries where Allen Carr's Easyway To Stop Smoking Centers are currently operational.

Check www.allencarr.com for latest additions to this list.

The success rate at the centers, based on the three-month money-back guarantee, is over 90 percent.

Selected centers also offer sessions that deal with alcohol, other drugs, and weight issues. Please check with your nearest center, listed below, for details.

Allen Carr's Easyway guarantees that you will find it easy to stop at the centers or your money back.

JOIN US!

Allen Carr's Easyway Centers have spread throughout the world with incredible speed and success. Our global franchise network now covers more than 150 cities in more than 45 countries. This amazing growth has been achieved entirely organically. Former addicts, just like you, were so impressed by the ease with which they stopped that they felt inspired to contact us to see how they could bring the method to their region.

If you feel the same, contact us for details on how to become an Allen Carr's Easyway To Stop Smoking or an Allen Carr's Easyway To Stop Drinking franchisee.

Email us at: join-us@allencarr.com including your full name, postal address, and region of interest.

SUPPORT US!

No, don't send us money!

You have achieved something really marvellous. Every time we hear of someone escaping from the sinking ship, we get a feeling of enormous satisfaction.

It would give us great pleasure to hear that you have freed yourself from the slavery of addiction so please visit the following web page where you can tell us of your success, inspire others to follow in your footsteps, and hear about ways you can help to spread the word.

www.allencarr.com/fanzone

You can "like" our facebook page here **www.facebook.com/AllenCarr**

Together, we can help further Allen Carr's mission: to cure the world of addiction.

CENTERS

LONDON CENTER AND WORLDWIDE HEAD OFFICE

Park House, 14 Pepys Road,
Raynes Park, London SW20 8NH
Tel: +44 (0)20 8944 7761
Fax: +44 (0)20 8944 8619
Email: mail@allencarr.com
Website: www.allencarr.com
Therapists: John Dicey, Colleen
Dwyer, Crispin Hay, Emma Hudson,
Rob Fielding, Sam Kelser, Sam Cleary,
Rob Groves, Debbie Brewer-West, Gerry
Williams (alcohol)

Worldwide Press Office

Contact: John Dicey
Tel: +44 (0)7970 88 44 52
Email: media@allencarr.com

UK Center Information and Central Booking Line

Tel: 0800 389 2115 (UK only)

CANADA

Sessions held throughout Canada
Email: mail@allencarr.com
Website: www.allencarr.com

USA

Tel: +1 855-440-3777
Sessions held throughout the USA
Email: mail@allencarr.com
Website: www.allencarr.com

Milwaukee (and South Wisconsin)

Tel: +1 262 770 1260
Therapist: Wayne Spaulding
Email: wayne@easywaywisconsin.com
Website: www.allencarr.com

UK CENTERS

Birmingham

Tel & Fax: 0800 389 2115
Therapists: John Dicey, Colleen
Dwyer, Crispin Hay, Emma Hudson,
Rob Fielding, Sam Kelser, Rob Groves,
Debbie Brewer-West, Gerry Williams (alcohol)
Email: mail@allencarr.com
Website: www.allencarr.com

Bournemouth

Tel: 0800 389 2115
Therapists: John Dicey, Colleen Dwyer,
Crispin Hay, Emma Hudson, Rob Fielding,
Sam Kelser, Rob Groves, Debbie Brewer-West
Email: mail@allencarr.com
Website: www.allencarr.com

Brentwood

Tel: 0800 389 2115
Therapists: John Dicey, Colleen Dwyer,
Crispin Hay, Emma Hudson, Rob Fielding,
Sam Kelser, Rob Groves, Debbie Brewer-West
Email: mail@allencarr.com
Website: www.allencarr.com

Brighton

Tel: 0800 389 2115
Therapists: John Dicey, Colleen Dwyer,
Crispin Hay, Emma Hudson, Rob Fielding,
Sam Kelser, Rob Groves, Debbie Brewer-West
Email: mail@allencarr.com
Website: www.allencarr.com

Bristol

Tel: 0800 389 2115
Therapists: John Dicey, Colleen Dwyer,
Crispin Hay, Emma Hudson, Rob Fielding,
Sam Kelser, Rob Groves, Debbie Brewer-West

Email: mail@allencarr.com
Website: www.allencarr.com

Cambridge

Tel: 0800 389 2115
Therapists: John Dicey, Colleen Dwyer,
Crispin Hay, Emma Hudson, Rob Fielding,
Sam Kelser, Rob Groves, Debbie Brewer-West
Email: mail@allencarr.com
Website: www.allencarr.com

Cardiff

Tel: 0800 389 2115
Therapists: Colleen Dwyer, Crispin Hay,
Emma Hudson, Rob Fielding, Sam Kelser,
Rob Groves, Debbie Brewer-West
Email: mail@allencarr.com
Website: www.allencarr.com

Coventry

Tel: 0800 321 3007
Therapist: Rob Fielding
Email: info@easywaymidlands.co.uk
Website: www.allencarr.com

Cumbria

Tel: 0800 077 6187
Therapist: Mark Keen
Email: mark@easywaymanchester.co.uk
Website: www.allencarr.com

Derby

Tel: 0800 389 2115
Therapist: John Dicey, Colleen Dwyer,
Crispin Hay, Emma Hudson, Rob Fielding,
Sam Kelser, Rob Groves, Debbie Brewer-
West
Email: mail@allencarr.com
Website: www.allencarr.com

Exeter

Tel: 0800 389 2115
Therapist: Colleen Dwyer, Crispin Hay,
Emma Hudson, Rob Fielding, Sam Kelser,
Rob Groves, Debbie Brewer-West
Email: mail@allencarr.com
Website: www.allencarr.com

Guernsey

Tel: 0800 077 6187
Therapist: Mark Keen
Email: mark@easywaymanchester.co.uk
Website: www.allencarr.com

Isle of Man

Tel: 0800 077 6187
Therapist: Mark Keen
Email: mark@easywaymanchester.co.uk
Website: www.allencarr.com

Jersey

Tel: 0800 077 6187
Therapist: Mark Keen
Email: mark@easywaymanchester.co.uk
Website: www.allencarr.com

Kent

Tel: 0800 389 2115
Therapists: John Dicey, Colleen Dwyer,
Crispin Hay, Emma Hudson, Rob Fielding,
Sam Kelser, Rob Groves, Debbie Brewer-
West
Email: mail@allencarr.com
Website: www.allencarr.com

Lancashire

Tel: 0800 077 6187
Therapist: Mark Keen
Email: mark@easywaymanchester.co.uk
Website: www.allencarr.com

Leeds
Tel: 0800 077 6187
Therapist: Mark Keen
Email: mark@easywaymanchester.co.uk
Website: www.allencarr.com

Leicester
Tel: 0800 321 3007
Therapist: Rob Fielding
Email: info@easywaymidlands.co.uk
Website: www.allencarr.com

Lincoln
Tel: 0800 321 3007
Therapist: Rob Fielding
Email: info@easywaymidlands.co.uk
Website: www.allencarr.com

Liverpool
Tel: 0800 077 6187
Therapist: Mark Keen
Email: mark@easywaymanchester.co.uk
Website: www.allencarr.com

Manchester
Tel: 0800 077 6187
Therapist: Mark Keen
Email: mark@easywaymanchester.co.uk
Website: www.allencarr.com

Manchester—alcohol sessions
Tel: +44 (0)7936 712942
Therapist: Mike Connolly
Email: info@stopdrinkingnorth.co.uk
Website: www.allencarr.com

Milton Keynes
Tel: 0800 389 2115
Therapists: John Dicey, Colleen Dwyer,
Crispin Hay, Emma Hudson, Rob Fielding,
Sam Kelser, Rob Groves, Debbie Brewer-West

Email: mail@allencarr.com
Website: www.allencarr.com

Newcastle/North East
Tel: 0800 077 6187
Therapist: Mark Keen
Email: mark@easywaymanchester.co.uk
Website: www.allencarr.com

Northern Ireland/Belfast
Tel: 0800 077 6187
Therapist: Mark Keen
Email: mark@easywaymanchester.co.uk
Website: www.allencarr.com

Nottingham
Tel: 0800 389 2115
Therapist: John Dicey, Colleen Dwyer,
Crispin Hay, Emma Hudson, Rob Fielding,
Sam Kelser, Rob Groves, Debbie Brewer-West
Email: mail@allencarr.com
Website: www.allencarr.com

Oxford
Tel: 0800 389 2115
Therapists: John Dicey, Colleen Dwyer,
Crispin Hay, Emma Hudson, Rob Fielding,
Sam Kelser, Rob Groves, Debbie Brewer-West
Email: mail@allencarr.com
Website: www.allencarr.com

Reading
Tel: 0800 389 2115
Therapists: John Dicey, Colleen Dwyer,
Crispin Hay, Emma Hudson, Rob Fielding,
Sam Kelser, Rob Groves, Debbie Brewer-West
Email: mail@allencarr.com
Website: www.allencarr.com

SCOTLAND
Glasgow and Edinburgh
Tel: +44 (0)131 449 7858
Therapists: Paul Melvin and Jim McCreadie
Email: info@easywayscotland.co.uk
Website: www.allencarr.com

Southampton
Tel: 0800 389 2115
Therapists: John Dicey, Colleen Dwyer,
Crispin Hay, Emma Hudson, Rob Fielding,
Sam Kelser, Rob Groves, Debbie Brewer-
West
Email: mail@allencarr.com
Website: www.allencarr.com

Southport
Tel: 0800 077 6187
Therapist: Mark Keen
Email: mark@easywaymanchester.co.uk
Website: www.allencarr.com

Staines/Heathrow
Tel: 0800 389 2115
Therapists: John Dicey, Colleen Dwyer,
Crispin Hay, Emma Hudson, Rob Fielding,
Sam Kelser, Rob Groves, Debbie Brewer-West
Email: mail@allencarr.com
Website: www.allencarr.com

Stevenage
Tel: 0800 389 2115
Therapists: John Dicey, Colleen Dwyer,
Crispin Hay, Emma Hudson, Rob Fielding,
Sam Kelser, Rob Groves, Debbie Brewer-West
Email: mail@allencarr.com
Website: www.allencarr.com

Stoke
Tel: 0800 389 2115
Therapist: John Dicey, Colleen Dwyer, Crispin
Hay, Emma Hudson, Rob Fielding, Sam

Kelser, Rob Groves, Debbie Brewer-West
Email: mail@allencarr.com
Website: www.allencarr.com

Surrey
Park House, 14 Pepys Road, Raynes Park,
London SW20 8NH
Tel: +44 (0)20 8944 7761
Fax: +44 (0)20 8944 8619
Therapists: John Dicey, Colleen
Dwyer, Crispin Hay, Emma Hudson,
Rob Fielding, Sam Kelser, Rob Groves, Sam
Cleary, Debbie Brewer-West, Gerry Williams
(alcohol)
Email: mail@allencarr.com
Website: www.allencarr.com

Watford
Tel: 0800 389 2115
Therapists: John Dicey, Colleen Dwyer,
Crispin Hay, Emma Hudson, Rob Fielding,
Sam Kelser, Rob Groves, Debbie Brewer-West
Email: mail@allencarr.com
Website: www.allencarr.com

Worcester
Tel: 0800 321 3007
Therapist: Rob Fielding
Email: info@easywaymidlands.co.uk
Website: www.allencarr.com

WORLDWIDE CENTERS

REPUBLIC OF IRELAND
Dublin and Cork
Lo-Call (From ROI)
1 890 ESYWAY (37 99 29)
Tel: +353 (0)1 499 9010 (4 lines)
Therapists: Brenda Sweeney and Team
Email: info@allencarr.ie
Website: www.allencarr.com

AUSTRALIA
ACT, NSW, NT, QSL, VIC
Tel: 1300 848 028
Therapist: Natalie Clays
Email: natalie@allencarr.com.au
Website: www.allencarr.com

South Australia
Tel: 1300 848 028
Therapist: Jaime Reed
Email: sa@allencarr.com.au
Website: www.allencarr.com

Western Australia
Tel: 1300 848 028
Therapist: Natalie Clays
Email: wa@allencarr.com.au
Website: www.allencarr.com

AUSTRIA
Sessions held throughout Austria
Freephone: 0800RAUCHEN
(0800 7282436)
Tel: +43 (0)3512 44755
Therapists: Erich Kellermann and Team
Email: info@allen-carr.at
Website: www.allencarr.com

BAHRAIN
Tel: 00966501306090
Website: www.allencarr.com

BELGIUM
Antwerp
Tel: +32 (0)3 281 6255
Fax: +32 (0)3 744 0608
Therapist: Dirk Nielandt
Email: info@allencarr.be
Website: www.allencarr.com

BRAZIL
Tel: +972 053 840-4080
Therapist: Lilian Brunstein
Email: lilian@asyaysp.com.br
Website: www.allencarr.com

BULGARIA
Tel: 0800 14104 / +359 899 88 99 07
Therapist: Rumyana Kostadinova
Email: rk@nepushaveche.com
Website: www.allencarr.com

CHILE
Tel: +56 2 4744587
Therapist: Claudia Sarmiento
Email: contacto@allencarr.cl
Website: www.allencarr.com

DENMARK
Sessions held throughout Denmark
Tel: +45 70267711
Therapist: Mette Fonss
Email: mette@easyway.dk
Website: www.allencarr.com

ESTONIA
Tel: +372 733 0044
Therapist: Henry Jakobson
Email: info@allencarr.ee
Website: www.allencarr.com

FINLAND
Tel: +358-(0)45 3544099
Therapist: Janne Ström
Email: info@allencarr.fi
Website: www.allencarr.com

FRANCE
Sessions held throughout France
Freephone: 0800 386387
Tel: +33 (4)91 33 54 55
Therapists: Erick Serre and Team

Email: info@allencarr.fr
Website: www.allencarr.com

GERMANY
Sessions held throughout Germany
Freephone: 08000RAUCHEN
(0800 07282436)
Tel: +49 (0) 8031 90190-0
Therapists: Erich Kellermann and Team
Email: info@allen-carr.de
Website: www.allencarr.com

GREECE
Sessions held throughout Greece
Tel: +30 210 5224087
Therapist: Panos Tzouras
Email: panos@allencarr.gr
Website: www.allencarr.com

GUATEMALA
Tel: +502 2362 0000
Therapist: Michelle Binford
Email: bienvenid@dejedefumarfacil.com
Website: www.allencarr.com

HONG KONG
Email: info@easywayhongkong.com
Website: www.allencarr.com

HUNGARY
Seminars in Budapest and
12 other cities across Hungary
Tel: 06 80 624 426 (freephone) or
+36 20 580 9244
Therapist: Gabor Szasz
Email: szasz.gabor@allencarr.hu
Website: www.allencarr.com

INDIA
Bangalore and Chennai
Tel: +91 (0)80 4154 0624
Therapist: Suresh Shottam

Email: info@easywaytostopsmoking.co.in
Website: www.allencarr.com

IRAN—opening 2019
Tehran and Mashhad
Website: www.allencarr.com

ISRAEL
Sessions held throughout Israel
Tel: +972 (0)3 6212525
Therapists: Ramy Romanovsky,
Orit Rozen
Email: info@allencarr.co.il
Website: www.allencarr.com

ITALY
Sessions held throughout Italy
Tel/Fax: +39 (0)2 7060 2438
Therapists: Francesca Cesati
and Team
Email: info@easywayitalia.com
Website: www.allencarr.com

JAPAN
Sessions held throughout Japan
www.allencarr.com

LEBANON
Tel: +961 1 791 5565
Therapist: Sadek El-Assaad
Email: info@AllenCarrEasyWay.me
Website: www.allencarr.com

MAURITIUS
Tel: +230 5727 5103
Therapist: Heidi Hoareau
Email: info@allencarr.mu
Website: www.allencarr.com

MEXICO

Sessions held throughout Mexico
Tel: +52 55 2623 0631
Therapists: Jorge Davo
Email: info@allencarr-mexico.com
Website: www.allencarr.com

NETHERLANDS

Sessions held throughout the Netherlands
Allen Carr's Easyway
'stoppen met roken'
Tel: (+31)53 478 43 62 /
(+31)900 786 77 37
Email: info@allencarr.nl
Website: www.allencarr.com

NEW ZEALAND
North Island – Auckland

Tel: +64 (0)27 4139 381
Therapist: Vickie Macrae
Email: vickie@easywaynz.co.nz
Website: www.allencarr.com

South Island – Dunedin and Invercargill

Tel: +64 (0)27 4139 381
Therapist: Debbie Kinder
Email: easywaysouth@icloud.com
Website: www.allencarr.com

NORWAY

Visit www.allencarr.com for latest
news of our services in Norway.

PERU
Lima

Tel: +511 637 7310
Therapist: Luis Loranca
Email: lloranca@dejardefumaraltoque.com
Website: www.allencarr.com

POLAND

Sessions held throughout Poland
Tel: +48 (0)22 621 36 11
Therapist: Anna Kabat
Email: info@allen-carr.pl
Website: www.allencarr.com

PORTUGAL
Oporto

Tel: +351 22 9958698
Therapist: Ria Slof
Email: info@comodeixardefumar.com
Website: www.allencarr.com

ROMANIA

Tel: +40 (0)7321 3 8383
Therapist: Diana Vasiliu
Email: raspunsuri@allencarr.ro
Website: www.allencarr.com

RUSSIA
Moscow

Tel: +7 495 644 64 26
Freecall +7 (800) 250 6622
Therapist: Alexander Fomin
Email: info@allencarr.ru
Website: www.allencarr.com

St Petersburg

Website: www.allencarr.com

SAUDI ARABIA

Tel: 00966501306090
Website: www.allencarr.com

SERBIA
Belgrade

Tel: +381 (0)11 308 8686
Email: office@allencarr.co.rs
Website: www.allencarr.com

SINGAPORE
Tel: +65 62241450
Therapist: Pam Oei
Email: pam@allencarr.com.sg
Website: www.allencarr.com

SLOVENIA
Tel: 00386 (0)40 77 61 77
Therapist: Grega Server
Email: easyway@easyway.si
Website: www.allencarr.com

SOUTH AFRICA
Sessions held throughout South Africa
National Booking Line:
0861 100 200 Head Office: 15 Draper Square,
Draper St, Claremont 7708, Cape Town, Cape
Town: Dr Charles Nel
Tel: +27 (0)21 851 5883
Mobile: 083 600 5555
Therapists: Dr Charles Nel,
Malcolm Robinson and Team
Email: easyway@allencarr.co.za
Website: www.allencarr.com

SOUTH KOREA
Seoul
Tel: +82 (0)70 4227 1862
Therapist: Yousung Cha
Email: master@allencarr.co.kr
Website: www.allencarr.com

SWEDEN
Tel: +46 70 695 6850
Therapists: Nina Ljungqvist,
Renée Johansson
Email: info@easyway.se
Website: www.allencarr.com

SWITZERLAND
Sessions held throughout Switzerland
Freephone: 0800RAUCHEN
(0800/728 2436)
Tel: +41 (0)52 383 3773
Fax: +41 (0)52 3833774
Therapists: Cyrill Argast and Team
For sessions in Suisse Romand
and Svizzera Italiana:
Tel: 0800 386 387
Email: info@allen-carr.ch
Website: www.allencarr.com

TURKEY
Sessions held throughout Turkey
Tel: +90 212 358 5307
Therapist: Emre Ustunucar
Email: info@allencarrturkiye.com
Website: www.allencarr.com

UNITED ARAB EMIRATES
Dubai and Abu Dhabi
Tel: +971 56 693 4000
Therapist: Sadek El-Assaad
Email: info@AllenCarrEasyWay.me
Website: www.allencarr.com

OTHER ALLEN CARR PUBLICATIONS

Allen Carr's revolutionary Easyway method is available in a wide variety of formats, including digitally as audiobooks and ebooks, and has been successfully applied to a broad range of subjects.
For more information about Easyway publications, please visit
shop.allencarr.com

Allen Carr's Quit Smoking Without Willpower

Allen Carr's Quit Smoking Boot Camp

Your Personal Stop Smoking Plan

The Illustrated Easy Way to Stop Smoking

Allen Carr's Easy Way for Women to Quit Smoking

The Illustrated Easy Way for Women to Stop Smoking

Finally Free!

Smoking Sucks (Parent Guide with 16 page pull-out comic)

The Little Book of Quitting Smoking

How to Be a Happy Nonsmoker

No More Ashtrays

The Only Way to Stop Smoking Permanently

Allen Carr's Quit Drinking Without Willpower

The Easy Way to Control Alcohol

Your Personal Stop Drinking Plan

The Illustrated Easy Way to Stop Drinking

Allen Carr's Easy Way for Women to Quit Drinking

No More Hangovers

The Easy Way to Mindfulness

Good Sugar Bad Sugar

The Easy Way to Quit Sugar

The Easy Way to Lose Weight

Allen Carr's Easy Way for Women to Lose Weight

No More Diets

The Easy Way to Stop Gambling

No More Gambling

No More Worrying

Get Out of Debt Now

No More Debt

No More Fear of Flying

The Easy Way to Quit Caffeine

Packing It In The Easy Way (the autobiography)

Want Easyway on your **smartphone** or **tablet**?
Search for "Allen Carr" in your app store.

Easyway publications are also available as **audiobooks**.
Visit **shop.allencarr.com** to find out more.

DISCOUNT VOUCHER
for
ALLEN CARR'S
EASYWAY CENTERS

Recover the price of this book when you attend an
Allen Carr's Easyway Center
anywhere in the world!

Allen Carr's Easyway has a global network of stop
smoking centers where we guarantee you'll find it easy
to stop smoking or your money back.

**The success rate based on this
unique money-back guarantee is over 90%.**

Sessions addressing weight, alcohol and other
drug addictions are also available at certain centers.

When you book your session, mention this
voucher and you'll receive a discount of
the price of this book. Contact your nearest
center for more information on how the sessions
work and to book your appointment.

**Details of Allen Carr's Easyway
Centers can be found at**
www.allencarr.com

This offer is not valid in conjunction with any other offer/promotion.